Math Girls^6

The Poincaré Conjecture

Hiroshi Yuki

Translated by Tony Gonzalez

http://bentobooks.com

MATH GIRLS 6: THE POINCARÉ CONJECTURE by Hiroshi Yuki

Originally published as *Sūgaku Gāru Powankare Yosō*

Softbank Creative Corp., Tokyo

Edited by M.D. Hendon and David Slutzky
Cover design by Kasia Bytnerowicz

Published 2022 by

Bento Books, Inc.
Houston, Texas 77043

bentobooks.com

ISBN 978-1-939326-49-2 (hardcover)
ISBN 978-1-939326-51-5 (trade paperback)
ISBN 978-1-939326-50-8 (case laminate)
Library of Congress Control Number: 2022931916

Printed in the United States of America
First edition, March 2022

Math Girls^6:

The Poincaré Conjecture

Contents

To my readers

This book contains math problems covering a wide range of difficulty. Some will be approachable by elementary school students, while others will challenge even college students.

The characters often use words and diagrams to express their thoughts, but in some places equations tell the tale. If you find yourself faced with math you don't understand, feel free to just browse through it before continuing with the story. Tetra and Yuri will be there to keep you company.

If you have some skill at mathematics, then please follow not only the story, but also the math. Doing so is the best way to fully find the shapes hidden within.

—Hiroshi Yuki

Prologue

> Another rarity is a person who throughout his life shows no shortcomings in terms of appearance, temper, or attitude.
>
> SEI SHONAGON
> *The Pillow Book*

Shapes, forms, contours . . .
These things seem obvious.
What you see is what it is.

Or is it?

Change a position, form changes too.
Change an angle, form changes too.
So how can we trust our eyes?
Sounds have form, scents have form, warmth has form.
And yet our eyes perceive none of these things.

Keys are small.
We can hold small things in a hand.
Space is large.
We can hold large things in ourselves.

Some forms are too small to be seen.

Others, too large.
Do we ourselves have form?

The small key in our grasp can open doors before us.
Through these doors, we can enter infinite space.

Perhaps this will allow me to find my own form.
Perhaps it will allow me to find yours.

CHAPTER 1

The Seven Bridges of Königsberg

> The branch of geometry that deals with magnitudes has been zealously studied throughout the past, but there is another branch that has been almost unknown up to now; Leibnitz spoke of it first, calling it the "geometry of position."
>
> LEONHARD EULER [10]

1.1 YURI

"Something's different about you recently," Yuri said.

It was a Saturday afternoon. I was in my room with my cousin Yuri, a third-year junior high student. We had played together since we were little, to the point where many people thought we were brother and sister. She was lying on the floor, flicking her chestnut brown ponytail while flipping through a stack of books she'd pulled off my shelves.

"Different how?" I asked.

Yuri turned a page. "Like, calmer. Even more boring than usual."

"More serious, I think you mean. That's just part of being a high school senior."

"Nah. We used to do all kinds of fun stuff together, but you've been super boring since summer vacation ended. And it's already fall!"

Yuri closed the book she'd been reading with a *smack*. It was a high-school math text, a fairly advanced one. I wondered how much of it she was able to understand.

"It's precisely because it's fall that I have to be so serious. I have college entrance exams coming up, you know. You've got your own entrance exams to get ready for, yeah?"

"Junior high students get serious in their own unique way."

I rolled my eyes. To be fair, though, I knew Yuri's grades were good enough for her to get into the high school she was aiming at, namely my school.

"Not that being serious makes school any more interesting." She sighed. I took this as meaning she still wasn't over her boyfriend, who had moved and was now attending a different school.

1.2 In a Single Stroke

"Hey, you ever heard of the Bridges of Königsberg?" I asked.

"The Bridges of... whattaberg?"

"Königsberg. It's the name of a town, one that had seven bridges."

"Sounds like you're about to tell me a fairytale, something about a brave knight having to cross seven holy bridges to face a fearsome dragon."

"Nothing like that. Actually, the Seven Bridges of Königsberg is a famous math problem."

"Surprise, surprise."

"Not one with equations, though—I guess you could call it a pathing problem."

"As in, finding a good path?"

"Sure. In this case, Königsberg has a river running though it and seven bridges over that river. Like this."

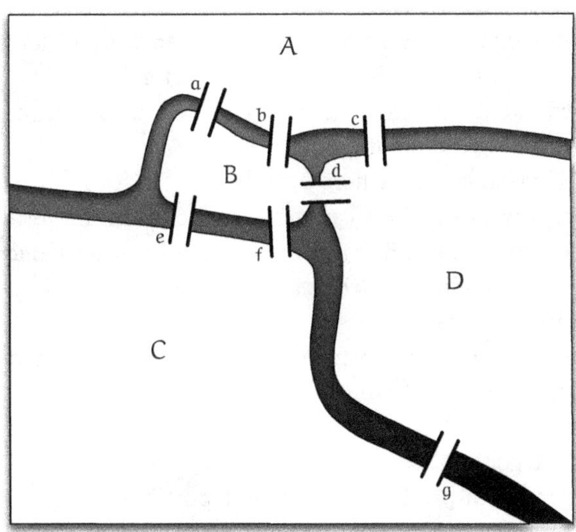

The Seven Bridges of Königsberg.

"There's only six bridges here, a through f," Yuri said.

"Looks like you missed g, here in the lower right corner. Also, these bridges connect four landmasses, A through D."

"Okay. And we're looking for a path over them?"

"Right. Specifically, a single path that will cross all seven bridges, but with an important condition—you can only cross each bridge once."

Problem 1-1 (The Seven Bridges of Königsberg)

Can you walk through Königsberg, crossing each of its seven bridges exactly once?

"I'm gonna bet there's actually another condition here."

I raised an eyebrow. "Which would be . . . ?"

"That I'm not allowed to swim across the river," Yuri said, grinning.

"Right, no swimming allowed. Your brave knight's armor is much too heavy. Bridges only."

"Oh, and another condition! I suppose I can't ask a friend to cross a bridge for me. Or six friends, for that matter."

I sighed. "Just one bridge-crosser. And no helicopters, no rockets, no tunnels, no 'beam me up, Scotty.' "

"Do I have to end up where I started?"

I shook my head. "You can start and end on any of the land-masses. All you have to do is cross each bridge exactly once."

"Well that should make things a lot easier."

"Give it a shot, then."

Yuri took a mechanical pencil from my desk and started drawing paths on the map I had sketched.

"Hmmm..."

"How's it going?"

"Not good. I'm pretty sure this isn't possible. Like, if I start on A, I can cross a → e → f → b → c → d, but then I'm stuck with no way to get to g."

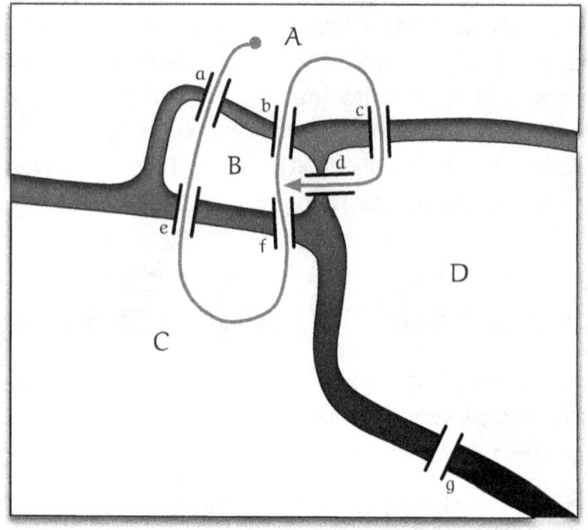

Crossing bridges a → e → f → b → c → d (g becomes unreachable).

"That's right," I said. "Once you cross bridge d to get to island B, you've used up all five bridges that would take you off of B, so you're stuck."

"Exactly."

"That's just one possible path, though. You're free to start from a different landmass, or take the bridges in a different order."

"I tried a bunch of ways, but none of them work!"

"Still, 'a bunch' isn't the same as trying all possibilities."

"Even so, I'm pretty sure it's impossible."

"So we can call that 'Yuri's conjecture.' "

"My what?"

"Conjecture. Through trial and error, you've come to the conclusion that there's no single path that crosses all bridges exactly once. You don't have a mathematical proof that you're correct, though. That's called a conjecture."

"A mathematical proof? Is something like that even possible? Doesn't seem like we can pull equations out of this, as much as you love them."

"Not a problem. There are ways of proving whether you can make a continuous walk through a graph without retracing any edges—which we call 'traversing' it—even without using equations."

"A graph? Edges?"

"Right. Not like the graph of a function, though, or a pie chart. A mathematical graph is a collection of vertices connected by what we call 'edges.' Investigating graph traversal is an important topic in math."

"But why are vertices and edges popping up all of a sudden?"

"Well, in this problem, we can think of the landmasses as being vertices, and the bridges as being edges. Then we get a graph that looks something like this."

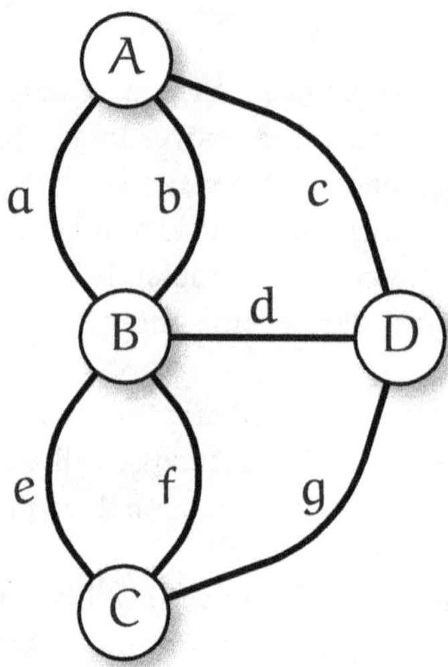

A graph for the Seven Bridges of Königsberg.

"In a graph, how the vertices are connected is very important," I said. "See how the connections are the same as in the map?"

"They look completely different to me."

"Not at all, I've just squeezed everything into a smaller space. See how the landmasses A through D in the map correspond to the vertices in the graph? And the bridges a through g correspond to the edges."

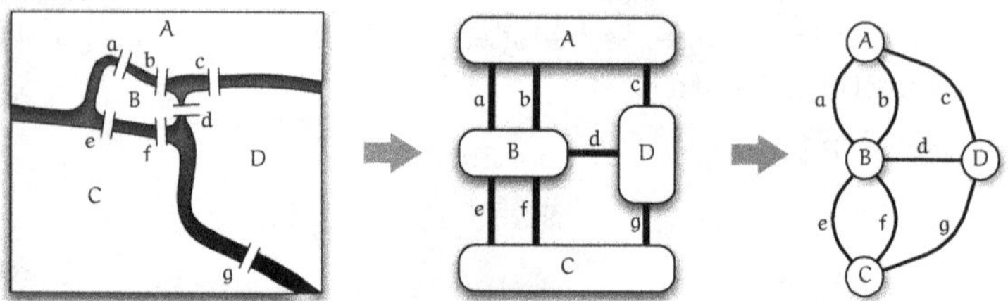

Squeezing the map down to make a graph.

"Okay, sure. I see it now."

"We're only interested in the path you take, not how big the landmasses are or how long the bridges are, so we can abstract all that geography out. The only thing that's important is which landmasses are connected by which bridges."

"Makes sense," Yuri said, nodding. "And it's okay for the bridges to be bent like this?"

"Sure. The shapes and lengths of edges don't matter either, so long as it's clear what they're connecting. Bridge g in the map is far from the others, but we can move it in closer, so long as we don't change any connections. The point of the graph is to better organize form. That makes proofs easier."

"How so?"

"Well, let's use this problem to see."

1.3 Starting Simple

"Let's start with something simpler," I said. "Say we're looking at this graph ①, with two vertices and one edge. This one's easy to traverse, right?"

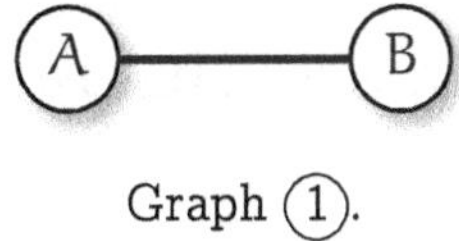

Graph ①.

"Well duh," Yuri said. "You just go from A to B and you're done."

"Right. Let's draw that with an arrow like this. We start at A and end at B, so we'll call A our starting point and B our endpoint."

Graph ① can be traversed.

"Fair enough."

"Okay, let's try this slightly more complicated graph ②, a triangle."

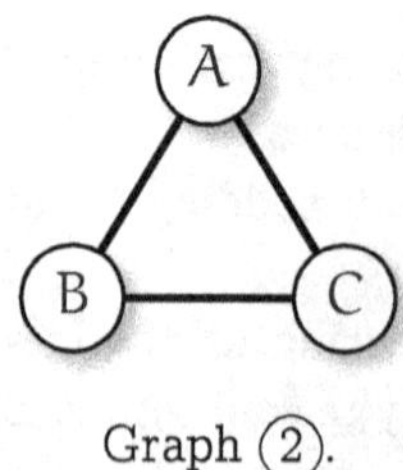

Graph ②.

"There's nothing complicated about this! We can just travel around it!"

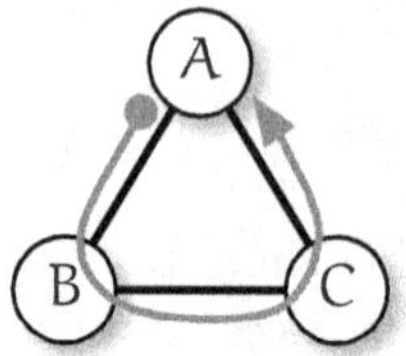

Graph ② can be traversed.

"Right. In this case, A is both the starting point and the endpoint."

"Yep, around and back."

"So how about this graph ③? Is it traversable?"

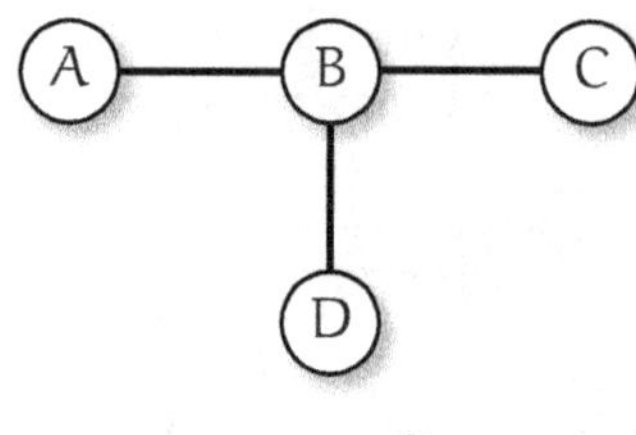

Graph ③.

"Nope!"

"Why not?"

"Well, just look at it. No matter where you start, you block yourself."

"That's right. If you choose A as your starting point, for example, you have to go to B next. Then you could go to C, and that leaves an edge to D, but you can't reach it. Why not?"

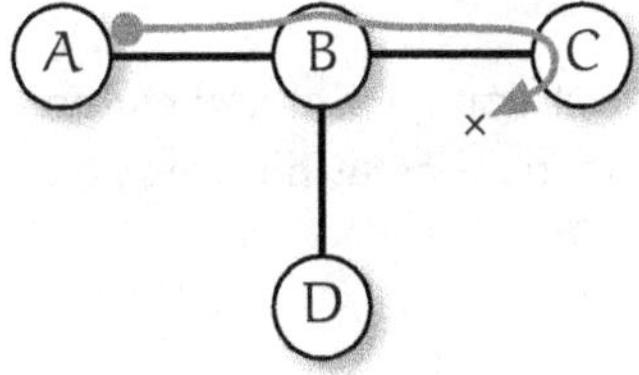

Graph ③ (with starting point A) cannot be traversed.

"Because we're stuck at vertex C."

"Right, because there's only one edge leading to vertex C, and we've already used that to get there. So yeah, we're stuck. Same thing for A → B → D, and something similar happens if we start from vertices C or D."

"Right."

"It's even worse if you start from vertex B. Then you can only cross one edge before you get stuck."

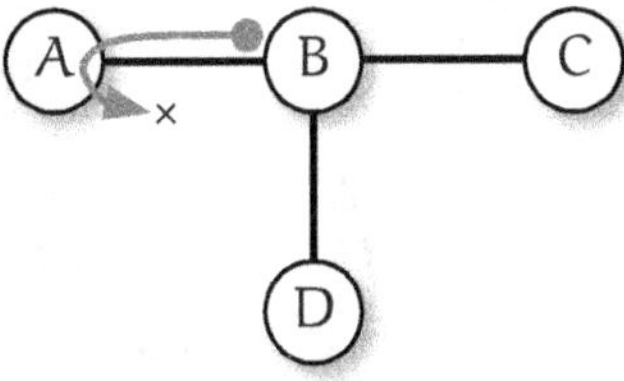

Graph ③ (with starting point B) cannot be traversed.

"Huh. So I guess we can't have any vertices with just one edge attached to them, because then we'll get stuck."

"Not so fast. That was the case with graph ③, but you can't generalize from just that. Remember graph ①? That had vertices with only one edge, and that was easy to traverse."

Vertices A and B in graph (1) have only one edge, but the graph can be traversed.

"Well those are the starting point and endpoint. That's different."

"Exactly! You've made a great discovery."

"I... I have?"

1.4 Graphs and Degrees

"You've made an important discovery about how to determine which graphs can be traversed," I said.

- We must consider how many edges connect to each vertex.
- We must separately consider starting points, endpoints, and intermediate points.

Yuri cocked her head. "Huh?"

"Say you have a traversable graph. Then the area around the starting point should look something like this, considering only edges connected to the starting point and ignoring what happens after you leave it."

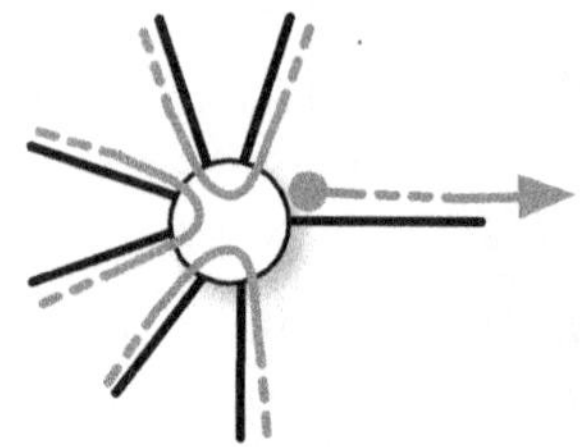

Starting point of a traversable graph.

"I'm not sure what's going on here," Yuri said.

"Look at the edges that connect to this starting point. In this case, there are seven. Since this is a starting point, one of those edges will be the one we used to leave it. Other edges will have to come in pairs, one to arrive at this vertex and one to leave it. There are three pairs like that in this graph, but there can be any number of pairs, even none."

"Okay, sure."

"So any starting point in a traversable graph will have one edge leaving it, along with some number of paired incoming and outgoing edges, which means it must have an odd number of connecting edges—$1, 3, 5, 7$, and so on."

"Very clever!"

"We can also say the same thing about the endpoint in such a graph."

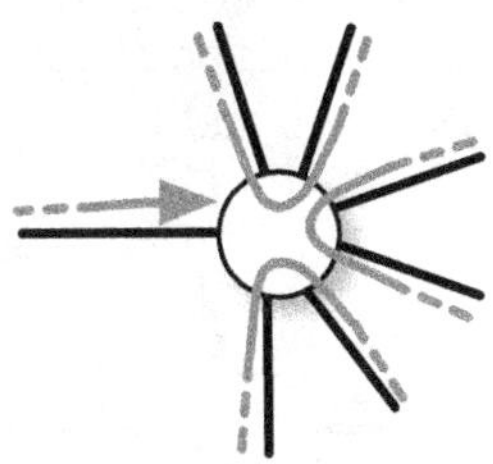

Endpoint of a traversable graph.

"So the endpoint will have an odd number of edges too," Yuri said.

"Yep, for the same reason. Any number of in–out pairs will have an even number of edges among them, and there will be one edge leading into the vertex, where we'll stop. Also, any vertices between the starting point and the endpoint will look like this."

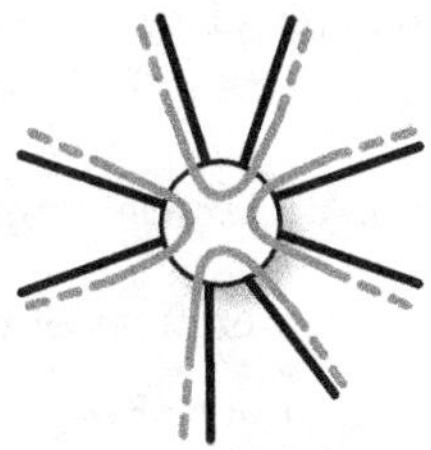

Intermediate point in a traversable graph.

"An even number!"

"That's right. It's an intermediate point, so incoming and outgoing edges will always come in pairs. So, an even number. There are only three kinds of vertices—starting points, endpoints, and intermediate points—so we've considered every possibility."

"Pretty cool!"

"One thing, though. We've only considered the case where the starting point and the endpoint are different vertices, but what do you think will happen if they're the same? In other words, the case where we start from one vertex, pass through all the others, and end up back where we started."

"Oh, I know! That combined start–end point will have an even number of edges, since the starting point and the endpoint will make a pair! The edge you leave from, and the one you come back from!"

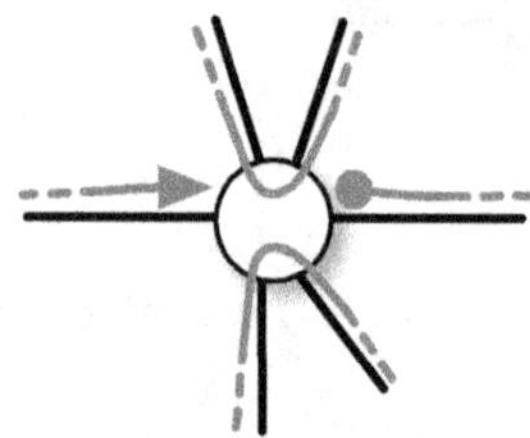

A graph where the starting point and the endpoint are the same vertex.

"That's right. When the starting point and the endpoint are the same, all vertices will have an even number of connecting edges. This is different from what we saw before, where starting points and endpoints have an odd number of edges, and all other points have an even number. Let's summarize everything we've seen so far."

- A traversable graph with *same* starting points and endpoints:
    - Starting point: Even number of edges
    - Endpoint: Even number of edges
    - Intermediate points: Even numbers of edges
- A traversable graph with *different* starting points and endpoints:

- Starting point: *Odd* number of edges
- Endpoint: *Odd* number of edges
- Intermediate points: Even numbers of edges

"Interesting..." Yuri said.

"So this brings up an important question."

> Given a traversable graph, how many vertices will have an odd number of edges?

"Well that's not hard. We just said there has to be either two or none at all. None if the starting point and the endpoint are the same, two if they're different. So... oh!"

"Noticed something?"

"The Bridges of Königsberg! It has four vertices with an odd number of edges!"

"That's right. A has 3 edges, B has 5, C has 3, and D has 3."

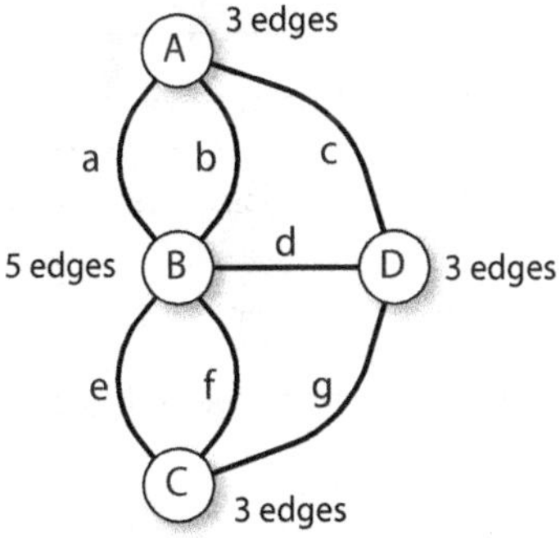

The graph for the Bridges of Königsberg problem has four vertices with an odd number of edges.

"So we can't traverse it!"

"That's right. Any traversable graph will have either zero or two vertices with an odd number of edges. But in the graph for the Bridges of Königsberg, there are four. So—"

"So we can't walk through the graph without recrossing somewhere!"

"No matter how we try, right. In other words, there is no solution to the Bridges of Königsberg problem, and we've completed our proof!"

Answer 1-1 (The Bridges of Königsberg)

If it were possible to traverse the Bridges of Königsberg, there would be either zero or two vertices with an odd number of edges. However, there are four such vertices in the graph for this problem. Therefore, the graph is not traversable.

"I like it," Yuri said, her eyes twinkling. "We can get a solution without performing a single trial."

"We call the number of edges connecting to a vertex the 'degree' of that vertex. So if we summarize what we learned about traversable graphs using that term instead, we get this."

What we've learned about traversable graphs

A traversable graph will have either zero or two vertices with an odd degree.

"Hey, kids! Teatime!" my mother shouted from downstairs.

1.5 Is this Math?

"My, you've grown," my mother said to Yuri.

"Ya think?" she replied, placing a hand on her head.

"I think she's in the middle of a growth spurt," I said.

We were in my living room, drinking the herbal tea my mother had served us. Well, Yuri was drinking it.

"Like it?" Mom asked.

"This is chamomile, isn't it?" Yuri said. "It's so comforting."

"You actually know what this stuff is?" I said.

"What do you think, dear?" Mom asked me.

"I'll let you know as soon as I've worked up the nerve to drink it," I said. Turning to Yuri, "So you're all good with the Bridges of Königsberg problem?"

"All good."

"Back to studying already?" My mother sighed and retreated to the kitchen.

"Euler was the mathematician who first gave a proof for that problem," I said. "Apparently he didn't consider it to be anything related to mathematics at first."

"I'd agree with him there."

"Not for long. He did end up finding the math of the problem. He even wrote a paper about its solution."

"What do you mean, he found the math?"

"The Bridges of Königsberg isn't just a simple logic problem, it's a math problem worthy of deep study. It's similar to geometry, in that it's a mathematical way of handling figures."

"Like, squares and circles?"

"Sure. But it's also different from normal geometry, in that we can freely change lengths and positions, so long as we don't change how things are connected."

"Oh, right. Like how you shrunk the graph way down before."

"Exactly. So long as things are linked up in the same way—so long as we retain connections, in other words—we can shrink large landmasses down to a single point, and we can make the bridges as short or as long or as bendy as we wish. So this Bridges of Königsberg problem planted a seed that grew into a new field of geometry."

"A new geometry, huh?"

"But when Euler wrote his paper in the eighteenth century, he didn't use a graph like we did. Those graphs didn't appear until the nineteenth century."

"But he did show we can prove things without any calculations?"

"Well, I wouldn't say there were no calculations at all. We did investigate even and odd degrees, after all. In his paper, Euler used a phrase from Leibniz—*geometria situs*, or 'the geometry of position.' Anyway, while Euler did plant the seed for this new form of geometry, it was Poincaré that fostered it and helped it grow. His papers used the phrase *analysis situs*, or 'analysis of position.' In the end, however, it came to be called 'topology.'"

"That sounds oddly familiar..."

"Topology is all about how things are connected. When we use geographical maps, it's important that everything on them is precisely located, right? But if we're just interested in how things are connected, precision in placement isn't so vital. So long as we don't change the *connections* between vertices and edges, we can place the vertices wherever we wish and we can stretch or shrink edges at will, since neither position nor length have anything to do with how we can choose paths. So what *is* important? Just the number of edges surrounding each vertex, in other words their degree, as you noticed. Quite brilliantly, I should add."

"Aw, cut it out. You'll make me blush."

"We found that in traversable graph problems, the number of vertices with odd degree is key. Hey, I know—let's call those 'odd points.' That will let us describe the conditions for those graphs more simply. We can just say, 'A traversable graph has zero or two odd points.'"

1.6 The "Opposite" Proof

"Euler didn't limit himself to just solving the Bridges of Königsberg problem," I said. "He also considered how to generalize solutions to similar problems. With a generalized solution, we can solve the Bridges of Königsberg problem more naturally. Of course, it's still important to use examples when thinking about a problem—it's hard to think generally when you've never thought specifically. To think slowly and carefully about a problem is also to measure your understanding of it. Examples are the key to understanding, after all.

"Then again, thinking only about a specific example isn't enough, either. When considering a concrete example, we should be looking to arrive at something more general. Euler ended his paper saying something like this."

> If the number of landmasses with an odd number
> of bridges—

- —is more than two, the desired journey is impossible.
- —is exactly two, the journey is possible if one starts in one of the odd numbered landmasses.
- —is zero, the journey is possible starting from any of the landmasses.

"Pretty much what we found, right?"

"Maybe..." Yuri said, "Or maybe it's the opposite?"

I blinked. "The opposite?"

"I mean, we ended up saying that if we can traverse the graph, it has zero or two odd points. But isn't Euler saying the opposite? That if a graph has zero or two odd points, we can traverse it?"

"Sure."

"And that's good enough?"

"Well, why wouldn't it be?"

"Because we haven't proved it that way yet! You just talked about the starting points, intermediate points, and endpoints in a traversable graph, right? We learned about graphs that we *can* traverse, but nothing about graphs that we *can't* traverse. So, like, maybe among all those graphs with zero or two odd points, there are some that *aren't* traversable. Could happen, right?"

"Well put!" I said.

Yuri sure is bright. She was exactly right. What we'd proved so far was something like this:

$$\text{Graph is traversable} \Longrightarrow \text{Graph has zero or two odd points}$$

But we hadn't proven the opposite:

$$\text{Graph is traversable} \Longleftarrow \text{Graph has zero or two odd points}$$

In other words, "If a graph has zero or two odd points, it is traversable" remained unproved.

"Mmm..." I groaned.

"Right? We haven't proved it, right? So prove it!"

I sat back to think. *Just how* can *we prove that?*

Yuri and I headed back to my room. I grabbed a sheet of paper off the stack on my desk, and started sketching a graph.

"Ah-ha!" Yuri said. "Look, I can create a graph with zero odd points that isn't traversable!"

"What? You've found a counterexample?"

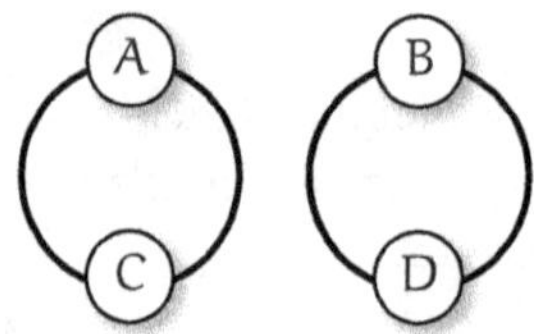

Graph (4) has zero odd points, but cannot be traversed.

"But that's . . . er, a perfectly valid counterexample, considering what we've said so far" I admitted. "Every vertex has a degree of 2, so there are no odd points. But still, this graph is split in two. It isn't connected, in other words. That's what makes it impossible to traverse."

"Yup."

"Okay, sure. You can't traverse a split graph. But that's kind of obvious, right? So I think we can reasonably limit ourselves to connected graphs, ones where no matter which two points we choose, there will be a path across some number of adjacent edges connecting them."

"I can live with that."

"While we're at it, we can't traverse a graph like this, either."

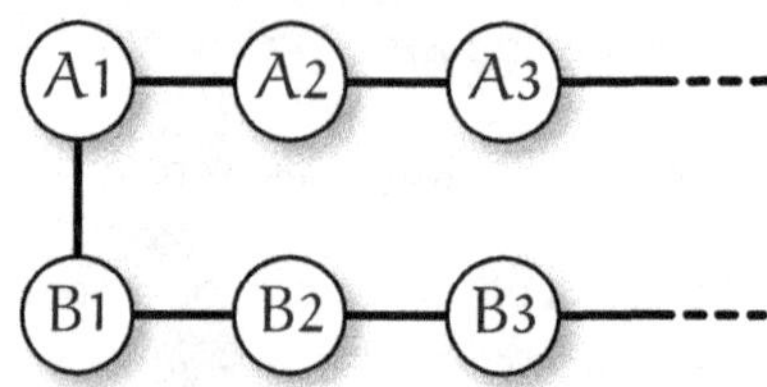

Graph (5) has zero odd points, but cannot be traversed.

"What's with the dashes heading off to the right?"

"I'm trying to show two infinite rows of vertices, $A_1, A_2, A_3, \ldots$ and $B_1, B_2, B_3, \ldots$."

"Wait, you can do that?"

"Well, here I just want to say we're going to exclude graphs like this when we talk about these traversal problems. In a graph like this one every vertex does have an even degree, but since it continues to infinity, there's no way to trace through the entire thing. So we want to limit our discussion to finite numbers of vertices."

"Hmmm..." Yuri thought for a moment. "In that case, we also need a condition saying we'll only have a finite number of edges."

"Well, if there's only finitely many vertices, it seems like there will only be finitely many edges..."

Yuri shook her head. "Bzzzt. We could make a graph like this."

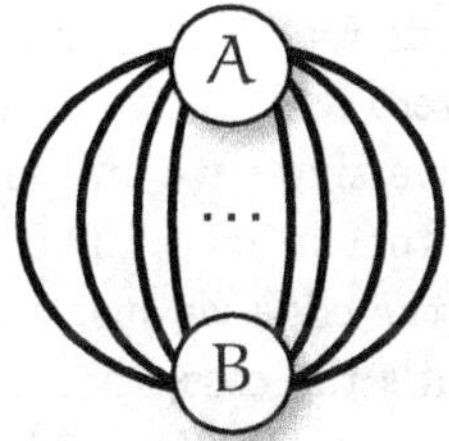

Graph ⑥ has finitely many vertices, but infinitely many edges.

"Ah, of course. Okay, you win. You could also say that in this graph, we can't determine the vertex's degree. Or that it's infinite, which isn't even or odd. Okay, so let's add conditions for finite numbers of vertices and edges."

Problem 1-2 (Reverse of problem 1-1)

If a connected graph with finitely many vertices and edges has zero or two vertices with odd degrees, is it traversable?

Yuri peered at my face. "These conditions are fine and all," she said, cocking her head and swinging her ponytail, "but is this really a hard problem?"

"Honestly, I'm not sure..."

I spent some time thinking and drawing graphs. Yuri sat next to me, working on her own. A good while passed as we worked through some good old trial-and-error.

"Okay, I think I've got it," I eventually said. "A way for finding a non-recrossing walk through a given graph with zero or two odd points. We can show it using a constructive proof."

"What's a constructive proof?"

"A proof that instead of showing that a traversal is possible, shows us how to create a traversal. Here, let's take things in turn.

"First off, let's clean up these separate cases of zero and two odd points. Say that if we're presented with the case where there are two odd points, we'll add a single edge connecting those vertices. Then the resulting graph is the case where there are zero odd points. So when considering methods for creating traversals, we only need to think about the case of zero odd points. Remember, if there are zero odd points, then our traversing path will always return to where it started from. We'll call that a 'loop.' The path through that loop will necessarily include the edge we just added, since it must pass through all edges. If that's the case, then if we remove that added edge we're back to the two-odd-points graph, and in a state where we could perform a traversal. So all we need to think about is traversal of a graph with zero odd points. In other words, just graphs of only even points. Good so far?"

"I think so. What's next?"

"Next is going back to what I mentioned about creating loops. That's very important."

"Important how?"

"Because my method for creating single-use paths is all about creating loops and connecting them."

"Wait, what?"

"Watch. We start by drawing a graph with only even points, and think about procedures for traversing it."

Procedure for traversing a graph

Consider a connected, finite graph with only even points and at least one edge.

- Start from any vertex and follow edges to create a loop L_1, then remove the edges in L_1 from the graph. Then, find a vertex in loop L_1 having an edge that has not yet been traveled.
- Starting from the found vertex, create a loop L_2, and remove its edges from the graph. Then, find a vertex in the loop formed by connecting L_1, L_2 having an edge that has not yet been traveled.
- Starting from the found vertex, create a loop L_3, and remove its edges from the graph. Then, find a vertex in the loop formed by connecting L_1, L_2, L_3 having an edge that has not yet been traveled.

$$\vdots$$

- Repeat this procedure until no vertex in the loop formed by connecting $L_1, L_2, L_3, \ldots, L_n$ has an edge that has not yet been traversed.

"Not really sure what's going on here," Yuri said, "but I guess it's something about making these random loops and connecting them? You sure something as simple as that works?"

"I'm sure. Here, let's follow along with a graph that has only even points."

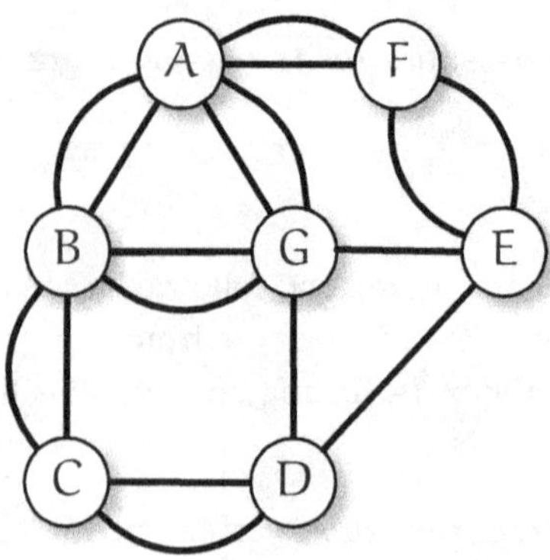

Finding a traversal in graph ⑦, which has only even points.

"So I just need to create a loop?" Yuri asked while drawing arrows connecting $A \to F \to E \to D \to C \to B \to A$.

"That's right. We'll call this one L_1."

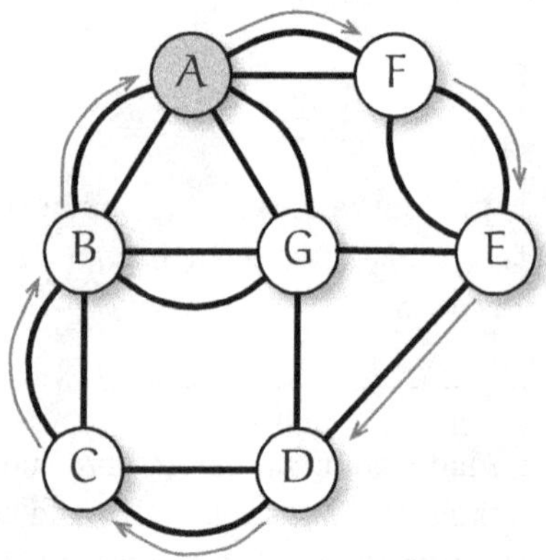

Create loop $A \to F \to E \to D \to C \to B \to A$ as L_1.

"Gotcha."

"Next, we remove all the edges in this loop L_1."

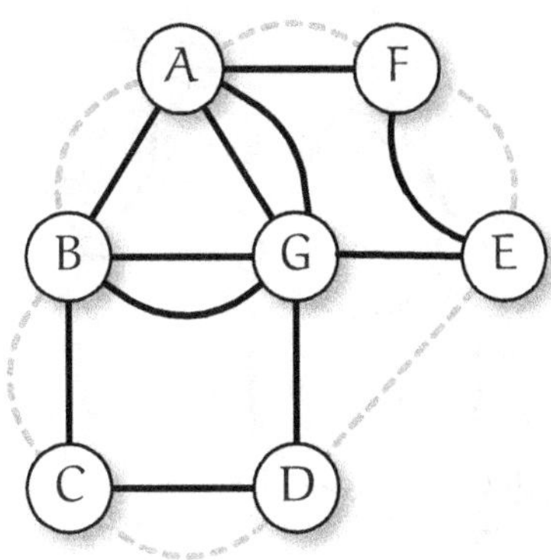

Remove the edges of L_1.

"Done."

"Great. See how even after removing all those edges, only even points remain? So we're going to keep creating and connecting loops to make a traversal."

"Hmm, okay. So now I should just make some loop other than L_1?"

"Yeah, but you have to think about which vertex you choose to start the loop. The rules say we have to choose one of the vertices we pass through in loop L_1 *that has an edge we haven't passed along yet.*"

"Not sure I get that."

"Well, loop L_1 goes A to F to E to D to C to B to A, right? We just want to choose one of those that still has an edge remaining after we removed L_1. So create a loop starting from, say, vertex F."

"I can do that!" Yuri said, tracing a loop $F \to A \to G \to E \to F$.

"Great, that will be L_2."

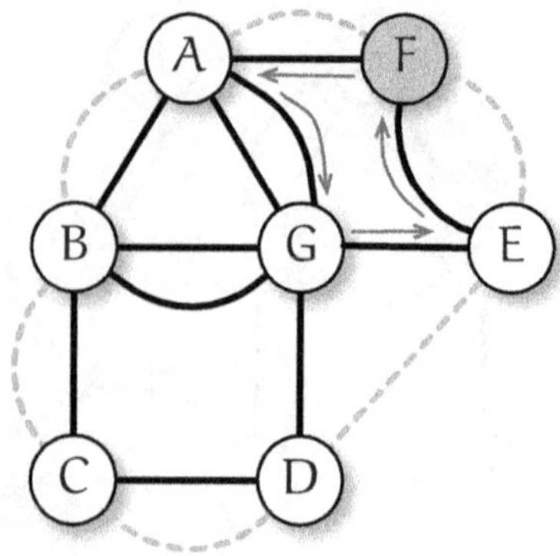

Create loop F $\to$ A $\to$ G $\to$ E $\to$ F as L_2.

"And now we remove the edges in L_2, right? Wow, our graph is getting pretty skimpy here."

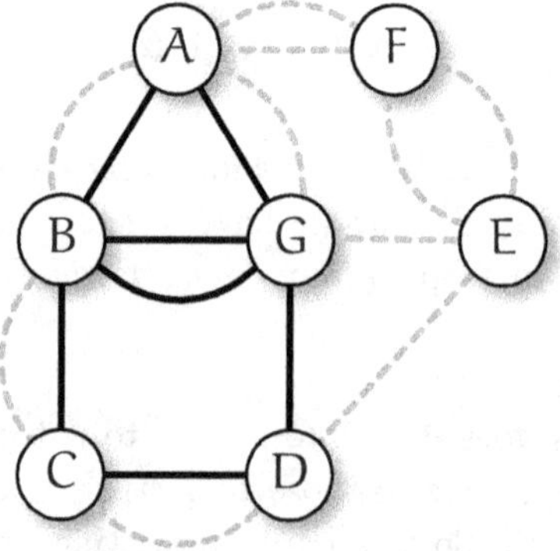

Remove the edges of L_2.

"It is. Now we connect L_2 to L_1, to create loop $\langle L_1, L_2 \rangle$."

"This connecting loops thing is where I'm still a little fuzzy."

"Vertex F acts as a connector between the two loops. Think of these as train stations and rail lines. As we're traveling along the L_1 line, when we arrive at station F we change tracks onto L_2. Then we loop around L_2, and when we come back to F we get back onto the

L_1 line. Then we just run along the rest of L_1. We can think of this as a new loop, $\langle L_1, L_2 \rangle$."

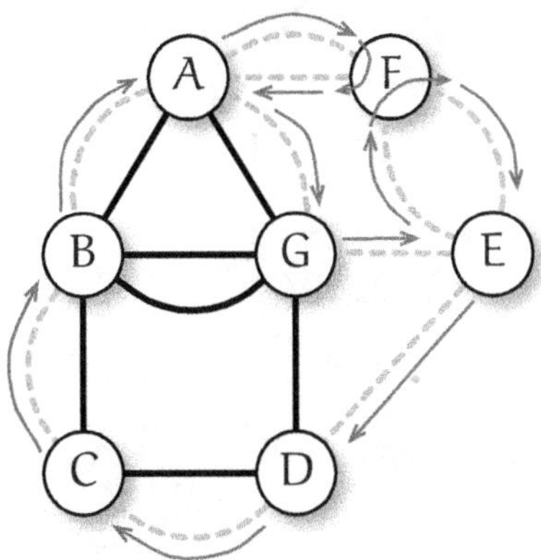

Create connected loop $\langle L_1, L_2 \rangle$.

L_1 A → F → E → D → C → B → A

Change tracks

Change tracks

L_2 F → A → G → E → F

"Oh, I get it! That's kinda cool!"

"Then it's just a matter of repeating all that. The vertex that starts loop L_3 will be one of those along loop $\langle L_1, L_2 \rangle$, one that still has an edge we haven't traveled along yet. Vertex A, for example."

"So I can start from A, and go $A \to B \to G \to A$ to make L_3?"

"You bet."

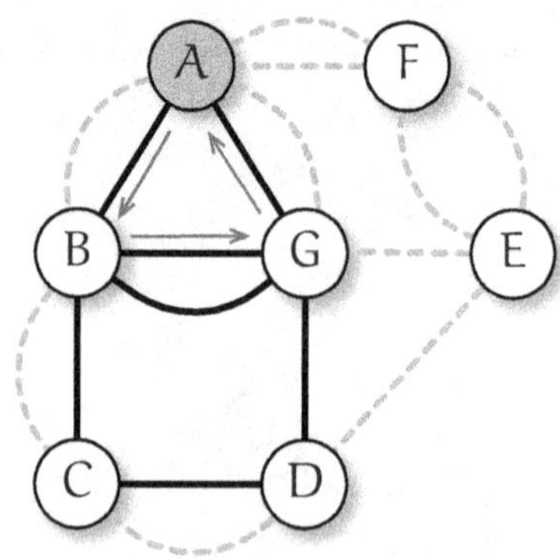

Create loop $A \to B \to G \to A$ as L_3.

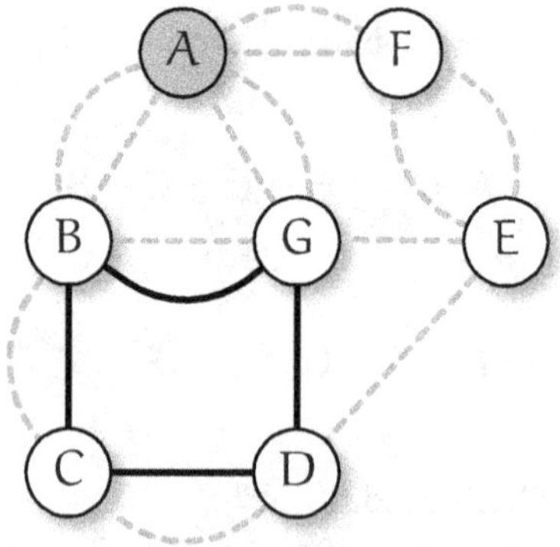

Remove the edges of L_3.

"Now we create $\langle L_1, L_2, L_3 \rangle$ like before, changing tracks at station, uh, vertex A."

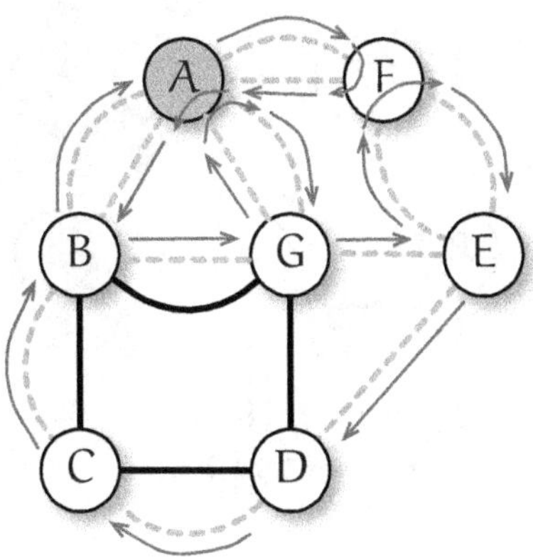

Create "big loop" $\langle\, L_1, L_2, L_3 \,\rangle$.

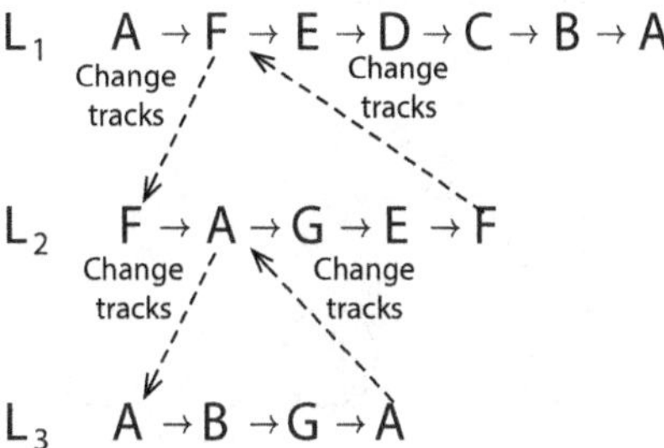

"So the next vertex is B, and . . . oh, it looks like the remaining edges form a loop."

"Right. We can let B → C → D → G → B be loop L_4, and remove those edges."

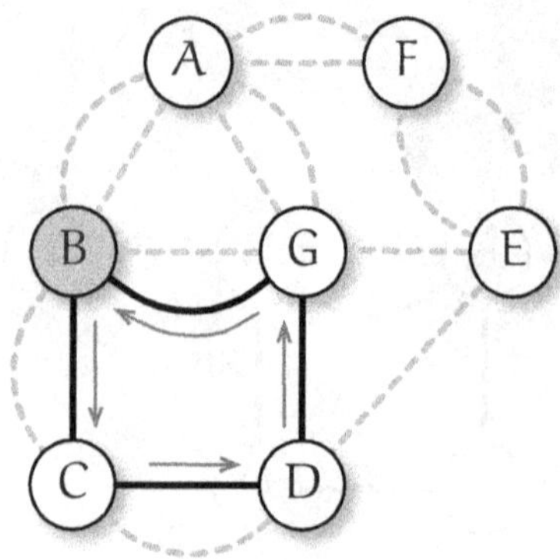

Create loop B → C → D → G → B as L_4.

"And that's it, they're all gone!"

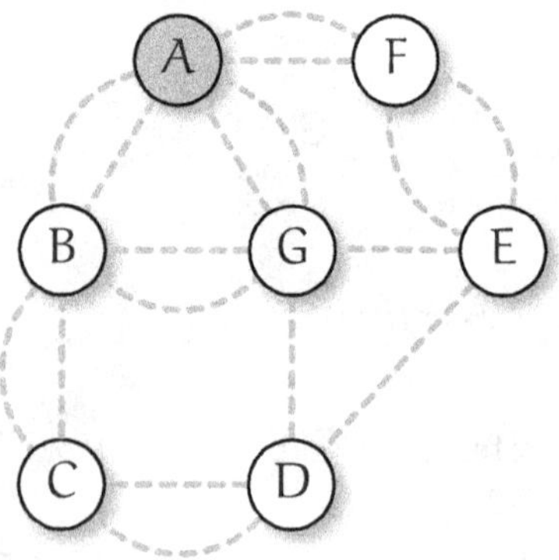

Remove the edges of L_4.

"So now that we have the connected loop $\langle L_1, L_2, L_3, L_4 \rangle$, we can traverse graph ⑦."

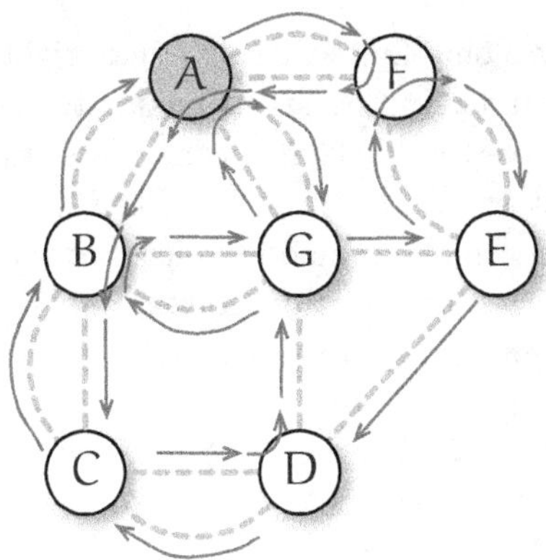

With the connected loop $\langle L_1, L_2, L_3, L_4 \rangle$, graph ⑦ can be traversed.

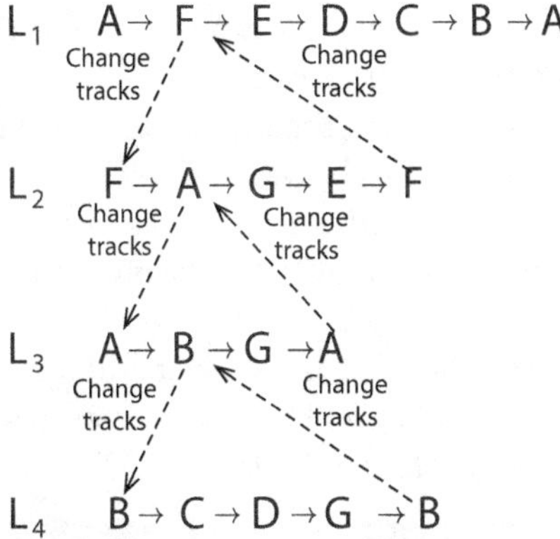

"Wow! That's... Wait, hold up. How do we know that didn't just happen to work for this particular graph? You haven't done anything to show this always works, you know."

"No, it will definitely work."

"How can you say that, when we're just, like, randomly selecting points? How do you know we can always make a loop?"

"Because of our conditions. We said the graphs we're thinking about will have finitely many edges, and that all vertices are even points. We know when we start from a given vertex and start following edges, we won't be doing that forever."

"Because there's only finitely many edges, right?"

"That's right. If we find ourselves unable to create a loop, we'll be stuck at some vertex X. That means there was some way into X but no way out, which would mean X had a way in but no way out. That would mean X was an odd point, but that contradicts our condition that all vertices are even points."

"Aha. So we *can* always create a loop."

"Indeed."

"Okay, I'm good. With that much at least—there's still some parts of your method that look suspicious. Like, we started from some vertex, made a loop, removed its edges, lather–rinse–repeat, but what says we can remove *all* the edges that way?"

"Another condition, that we're only considering connected graphs."

"Well yeah, I know the graph started out connected, but we're yanking out all these edges, right? Seems like that could break the original graph up into several pieces."

"It could, sure. But those separated bits—'connected components,' let's call them—will always have some vertex in common with a previously constructed loop. If that weren't the case, the graph wouldn't have been connected to begin with."

"Huh, okay."

"So anyway, this method shows the reverse of what we had before. It's interesting that simple numbers like vertex degrees tell us so much about whether graphs can be traversed."

Answer 1-2 (Reverse of problem 1-1)

If a connected graph with finitely many vertices and edges has zero or two vertices with an odd degree, it is traversable.

"And having settled that issue . . . I'm hungry."

"Are you kidding? We just had snacks!"

"No, we had chamomile tea," Yuri said, grinning. "That's not enough for a growing girl."

She skipped out of my room, leaving me alone to think.

It really is *interesting that simple numbers like vertex degrees tell us so much about graphs. From those numbers alone, we can judge whether they can be traversed. But even so...*

I looked at the textbooks on my desk, sitting next to a schedule I'd made of the days leading up to my college entrance exams.

Even so... I still don't have a good way of judging what I should do after high school. After I take my exams, will those scores tell me what I should do? No, just whether I can get into college. Entrance exams are an intermediate point, not an ending point. What meaning can they have for me, or my future? I—

"Yo, dude!" At the sound of Yuri's voice, I snapped back to reality; it had an insistence to it I'd never heard before. "Get down here! Quick!"

I hurried down the stairs, then dashed through our living room and into the kitchen.

Where I found my mother lying on the floor.

"M–Mom?"

> "If there are more than two regions which are approached by an odd number of bridges, no route satisfying the required conditions can be found. If, however, there are only two regions with an odd number of approach bridges, the required journey can be completed provided it originates in one of the regions. If, finally, there is no region with an odd number of approach bridges, the required journey can be effected, no matter where it begins."
>
> LEONHARD EULER
> *Solutio problematis ad geometriam situs pertinentis* [10]

CHAPTER 2

Möbius Strips and Klein Bottles

> That's right—foam. An infinitude of fine bubbles. The form they take is so interesting, I've found myself just sitting there, staring at them.
>
> HIROSHI MORI
> *Sky Eclipse*

2.1 ON THE ROOF

2.1.1 Tetra

"Oh, that's awful!" Tetra said.

"Thankfully it wasn't anything serious," I replied. "She says she just got dizzy and fell. She's going to the doctor to get it checked out, just in case."

It was lunchtime, and we were up on the school's roof. The wind was gentle, but had a slight chill. The sycamores in the schoolyard had already lost most of their leaves. *Sure enough, fall has arrived.*

I had told Tetra about my mother's condition as I ate a bun I'd bought at the school store. Seeing my mother collapsed on our kitchen floor had been a shocking experience. Equally surprising had been when she stood up under her own power and gave a self-conscious chuckle. She seemed to be okay, but still...

"Well I'm glad to hear it was nothing serious!" Tetra said, returning to the fried egg in her bento box.

"I guess..." I said, wondering what "serious" might have been. I didn't say anything to Tetra, but I hadn't been able to shake a feeling of dread ever since. My mother had always been so healthy—I'd never seen her with anything more serious than a cold—so seeing her totally incapacitated had hit me hard. I couldn't imagine anything more worrying than a sick parent.

I decided to change the subject. "Studying any fun math recently?"

Tetra was a second year student, one year behind me. When she'd first come to our school she'd been struggling with math, but now she was really enjoying it. She always seemed to be plugging away on some interesting problem.

"Nothing in particular right now," she said, "but ever since my presentation at the Narabikura Library[1] and attending the Galois Festival there,[2] there *is* something I've been kinda thinking about..."

"Oh yeah? What's that?"

"Actually, no. Never mind. It's still a secret." Tetra blushed and pressed a hand to her mouth.

2.1.2 Möbius Strips

Finishing her lunch, Tetra packed away her bento box, wrapping it in a pink cloth.

"By the way," she said, "have you ever heard of Möbius strips?"

"Sure," I said.

"I saw something about them on a TV show last night. You take a strip of paper and, like, give it a twist..." Tetra mimed constructing a Möbius strip.

"Yep. It looks like this, right?" I pulled out a notebook and started sketching. "Actually you don't give it a twist, you give it a half twist."

---

[1]See *Math Girls 4: Randomized Algorithms.*

[2]See *Math Girls 5: Galois Theory.*

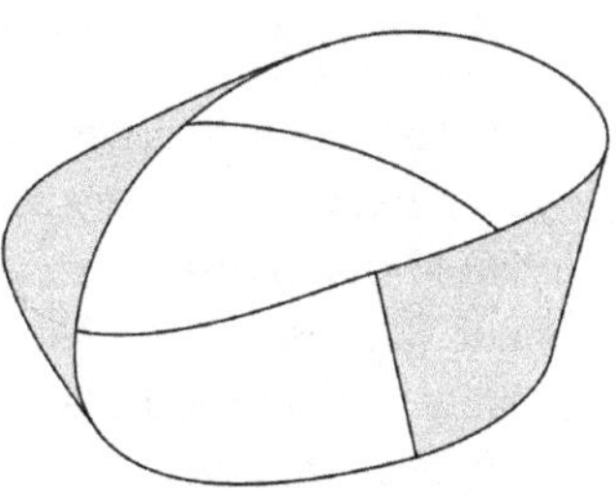

A Möbius strip.

"A half twist, right! So what's the big deal about a half-twisted strip of paper? Why does it have a name and everything? Möbius was a mathematician, right? Is there something, like, mathematically important about this shape? They didn't really explain all that on the TV show."

Good old Tetra, always grasping the essential, fundamental questions, and never pretending she knows the answers. Always asking herself if she truly understands what's going on.

"There's a lot that's interesting about Möbius strips," I said. "For example, if you create a loop from a strip of paper without twisting it, you get a cylinder."

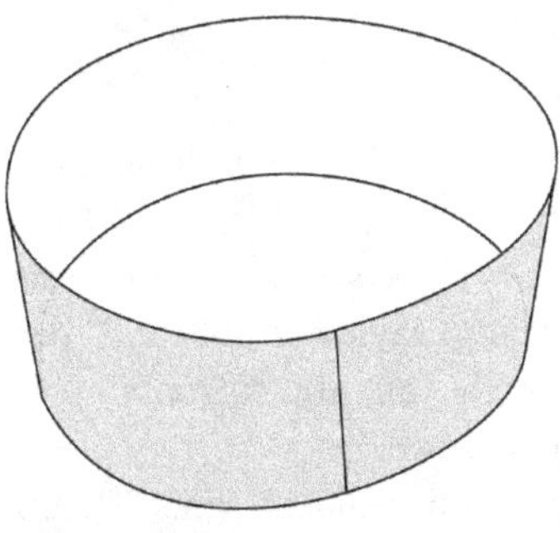

A cylinder.

Tetra nodded. "Right."

"So what happens if you color one side of the cylinder? Say you paint the outside red. Then only the outside gets colored, and the inside remains untouched, right?"

"Well sure."

"Then you could paint the inside a different color—blue, for example. Anyway, then you'd have the inside and the outside painted different colors, so you could distinguish between them."

"And I guess that's not the case with a Möbius strip?"

"That's right. The only difference between a cylinder and a Möbius strip is that half twist, but you don't get the same result when you try to paint one face of a Möbius strip. If you follow the face of a Möbius strip, painting as you go, you end up back where you started. You end up with the entire thing painted. So you can't distinguish between the front and back of a Möbius strip."

"I guess so, yeah. When you're painting one side that half twist takes you to the other, so you end up with red on both the front and the back!"

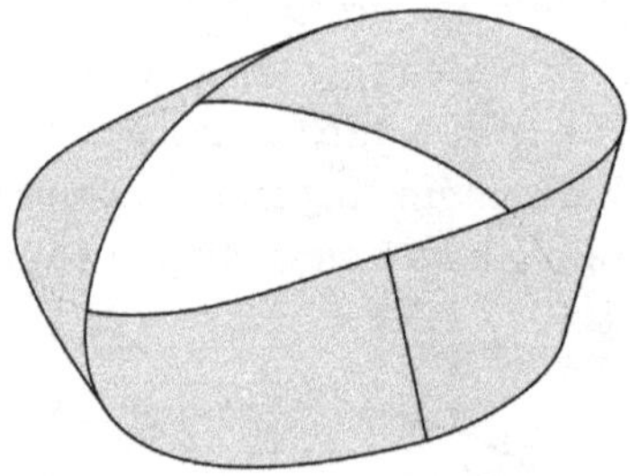

A painted Möbius strip.

"Right, except that since you've painted one side and ended up coloring the whole thing, saying 'front' and 'back' isn't really accurate. You can only really talk about fronts and backs when you can distinguish between the two."

"Yeah, I guess so. But still—so what?" Tetra said this looking straight at me with her charming large eyes.

There it is. Tetra's questions were always straightforward and to the point.

I had encountered Möbius strips in many books. They're a pretty popular mathematical topic, after all. But I'd never stopped to think "So what?" I considered the Möbius strip to be an interesting shape, and I knew how you couldn't distinguish between their front and back. But why do they get so much attention? That's something I'd never considered.

Tetra peered at me, put off by my long silence. "Sorry, I guess that was a weird question?"

"No, no need to apologize. It's a great question, actually. I'm just not sure how to answer it. I know Möbius strips are cylinders with a half twist, and I know they have something to do with a branch of mathematics called topology, but why are they so important? I'm not really sure."

"Oh, okay."

The warning bell rang, so we packed up and headed back to our respective classrooms, each with nagging questions about half twists.

2.2 In the Classroom

2.2.1 Independent Study

My next period was a study hall, which I'd decided I would spend reviewing some physics problems.

Since I was in my third and final year at a college preparatory high school, much of my fall quarter schedule was filled with elective "independent study" courses, which allowed students to prepare for their college entrance exams however they thought best. Each student had different strengths and weaknesses, and we were aiming at different schools with their own entrance exams of varying difficulty, so a one-size-fits-all schedule just wouldn't work. Indeed, looking around the room I saw classmates studying pretty much every subject. Each to their own, according to the paths they'd chosen for themselves.

After working through a few physics problems, I paused to once again consider what path *I* should choose. Honestly, I was still unsure. I didn't know what I was good at, or what I might accomplish. I didn't know anything about the world beyond school. I didn't even really know much about myself.

Every time an adult asked me what I wanted to do in the future, I would cringe. Among all the questions about the decisions facing me, that was the worst. I didn't mind being asked if I would pick a major in the sciences or something in the humanities (a math or science major, obviously), and I didn't mind being asked what my strengths were (math, obviously), or even what I wasn't so good at (geography and history). When guidance counselors asked me what schools I intended to apply to, I could answer that too. I had to list my top three, in fact, when I applied for practice tests to see if I was making reasonable choices.

Being asked what I actually wanted to do in the future, though? That I didn't like, because I didn't have an answer. I hated not having an answer. There was something... *irritating* about not knowing what I was supposed to be aiming for, like my own "form" was unclear. I felt spineless and squishy, like slime.

So what do *I want to do?*

2.3 In the Library

2.3.1 Miruka

"Because mathematicians have an interest in sameness," Miruka said.

"They're interested in... *sameness*?" Tetra replied.

"Yes, sameness."

We were in the school library after classes, Tetra and I sitting at a table across from Miruka, my classmate. She had long, black hair, metal-framed glasses, and a genius for mathematics. The three of us often gathered in the library after classes to talk about math.

Tetra had asked Miruka the question I'd been unable to answer—why Möbius strips were so important—causing Miruka to grin and enter math mode.

"When we study any mathematical object—numbers, shapes, functions, whatever—we always pay attention to what's the same as what. Mathematicians prefer precision in their discussions. After all, those discussions won't make any sense if it isn't clear what we're talking about, and that includes whether two things we're presented with are different or the same. We won't make much progress if we can't even determine that, right?"

"Though in the case of numbers, we would say 'equal' instead of 'same,' right?" I said.

"Except that I'm speaking at a more abstract level," Miruka said. " 'Equal' is just one kind of 'same.' We can easily define other kinds of sameness."

"Other kinds of sameness?" Tetra said. "Like, a way to say that 1 and 7 are the same?"

"Let's see..." Miruka began speaking a little slower. "Well, we know the concept of parity, right? Whether a number is even or odd. 1 and 7 are both odd, so they have the same parity. In that way we can use parity to talk about the sameness of two numbers."

"The point being that there are many kinds of sameness, right?" I said.

"Exactly." Miruka pushed her glasses back with her index finger. "Back to Möbius strips. Tetra, you described them as being 'a cylinder with a half twist,' right?"

"I did!"

"If you think about it, though, it isn't really a question of whether a half twist *exists*, is it? The question is, *how many* half twists are there? In particular, whether there are an odd number of them."

"Oh, of course!" I said. "If there are an even number of half twists, we're back to being a cylinder, aren't we? If the number of half twists is $0, 2, 4, 6, \ldots$, in other words an even number, then we can distinguish one side from the other, just like with a cylinder. Zero half twists is a normal cylinder. Make an even number of half twists and you don't have a Möbius strip, even though it kind of looks like one."

Tetra nodded. "Sure, that makes sense. And if you make $1, 3, 5, 7, \ldots$ half twists, an odd number of them, you can't distinguish front from back."

"You haven't mentioned negative numbers yet," Miruka prodded.

"Negative numbers?" I said. "Oh, okay. I guess that would correspond to twisting in the opposite direction."

Categorization by number of half twists (distinguishing front from back)

- After applying an even number $(\ldots, -4, -2, 0, 2, 4, \ldots)$ of half twists:
  A curved surface for which we can distinguish front from back ("same" as a cylinder)

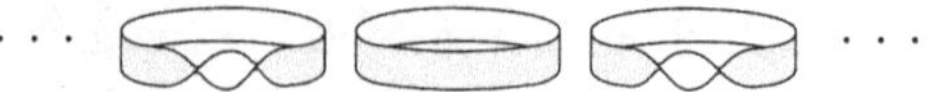

- After applying an odd number $(\ldots, -5, -3, -1, 1, 3, 5, \ldots)$ of half twists:
  A curved surface for which we cannot distinguish front from back ("same" as a Möbius strip)

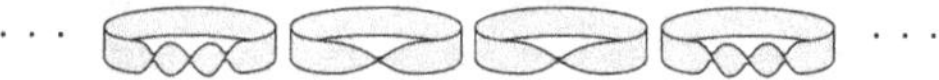

"So to state the obvious," I said, "there's a correspondence between the parity of the number of half twists and whether we can distinguish between the front and back of the surface."

"Simple though it may be, it's an example of a correspondence between a property of numbers and a property of figures."

Tetra raised her hand, as if we were in class. "I have a question."

"Yes, Tetra?" Miruka said, pointing at Tetra to faithfully play the role of teacher.

"Sorry to interrupt, but there's something I still don't get. How to put this... Um, I understand how we're using half twists to create cylinders and Möbius strips, and I see how we can make two categorizations of strips according to whether they have an even or odd number of half twists. But, is that really important? I mean, I sort of feel like it must be, but if you asked me to say specifically *why* it's important? Well, I don't think I understand this well enough to answer in my own words."

"Hmph." Miruka closed her eyes and pressed a finger to her lips as she thought. Tetra and I held our breath, waiting to hear what she would say.

Finally: "Time for step one in our study of classification."

2.3.2 Classification

"When we're presented with various objects," Miruka began, "we instinctually want to categorize them. This isn't limited to mathematical objects. We do the same with animals, and plants, and minerals, right? In general, that's called natural history. When performing these categorizations, the basis for our judgment is whether one thing and another are the same or different.

"As a math example, say we have the set of all integers $\ldots, -3, -2, -1, 0, 1, 2, 3, \ldots$, and we want to sort it into two categories, even and odd. Mathematically, we would represent the set of all integers as a union of sets having no common elements, a categorization with no leaks and no dupes. We call putting things in categories like this *classification*."

Classify the set of all integers according to parity.

$$\text{Set of all integers} = \{\ldots, -4, -3, -2, -1, 0, 1, 2, 3, 4, \ldots\}$$

$$\Downarrow$$

$$\text{Set of even integers} = \{\ldots, -4, -2, 0, 2, 4, \ldots\}$$

$$\text{Set of odd integers} = \{\ldots, -3, -1, 1, 3, \ldots\}$$

$$\{\text{Even integers}\} \cup \{\text{Odd integers}\} = \{\text{All integers}\}$$

$$\{\text{Even integers}\} \cap \{\text{Odd integers}\} = \{\ \} = \text{Empty set}$$

"In this classification, the standard we used was whether each integer leaves a remainder of 0 or 1 after being divided by 2. If the remainder is 0, the integer is even. If 1, odd. By doing so, every integer is classified as even or odd, so there are no leaks. Further, no integer is classified as both even and odd, so there are no dupes. There you go, a perfect classification.

"We can create a similar classification for the set of all shapes resulting from half twists in cylinders. In this case, we use as our criterion whether we can distinguish between the front and back of the surface. Elements in the 'set of all shapes resulting from repeating half twists of a cylinder' will be either 'a shape with distinguishable front and back' or 'a shape with indistinguishable front and back.' This too is a classification with no leaks and no dupes, according to the parity of the number of half twists.

"So here's where things get serious. Why is the Möbius strip an important shape? Because this criterion, the distinguishability of front and back surfaces, is mathematically important. Specifically, it's related to the classification of closed surfaces, a problem in topology that was solved in the nineteenth century. Infinitely many closed surfaces are possible, so how can we put them into categories with no leaks and no dupes? In other words, how can we classify them? Well, the first step in that direction is to figure out how we can compare two of those surfaces, so we can say whether they're the same or different.

"Of course, an extremely simple classification would be to ask whether two surfaces are the same in all ways, or if they're different in all ways. Those are pretty clear criteria, sure, but they aren't very useful. A good classification provides a clear view of what we're studying. We've only really advanced the field when through the process of determining what's the same and what's different, we've obtained a *useful* standard.

"In mathematics, this property of being able to distinguish between front and back is called 'orientability,' and this is an important standard for classifying closed surfaces."

2.3.3 Classification of Closed Surfaces

"Interesting," I said. "So you're saying we can classify closed surfaces according to their orientability?"

Tetra held up a hand. "Whoa, back up a minute. Before we get into that . . . what's a closed surface?"

"Simply put, a closed surface is one that doesn't have any boundaries, but doesn't spread out to infinity. For example, cylinders and Möbius strips have boundaries, so neither are closed surfaces."

"They have boundaries?" Tetra asked. "How?"

"A cylinder has two boundaries," Miruka said, "a Möbius strip has just one." She made some sketches with emphasized borders and numbered them.

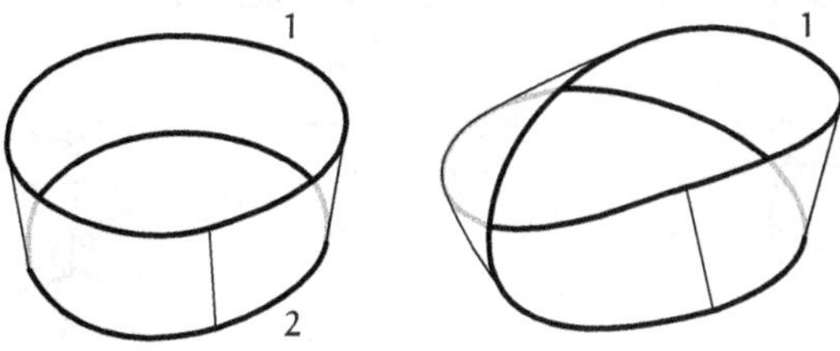

"Oh, so they're like edges!" Tetra said. "How cool, that a Möbius strip only has one!" She traced her finger around and around, following it.

"One big loop," I said.

2.3.4 Orientable Surfaces

"We said a closed surface is one that has no boundaries, but still doesn't extend to infinity," Miruka said. "There are precise mathematical definitions for those conditions, but for now let's just look at some examples. A sphere is a good one to start with."

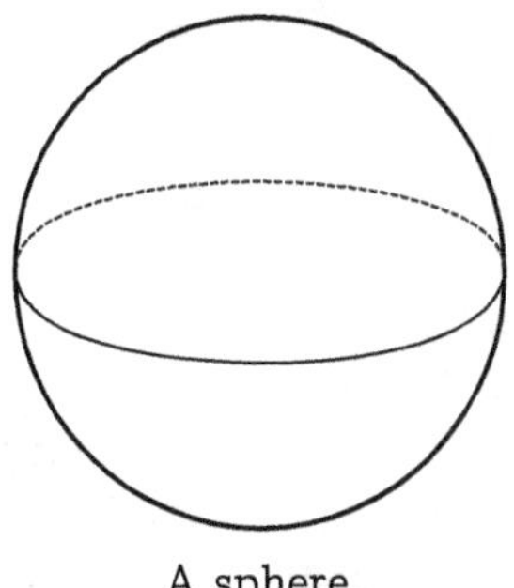

A sphere.

"Like, a ball, right?" Tetra said.

"Except that when we use the word 'sphere' in mathematics we're only considering a surface. There's nothing in a sphere's interior. A 'ball' in math is that surface filled with stuff, a solid."

"Ah, okay," Tetra said, nodding.

"When we think about classifications of closed surfaces in topology, we consider distortions like stretching or shrinking to leave shapes fundamentally unchanged. So topologically speaking, each of these closed surfaces is the 'same' as a sphere."

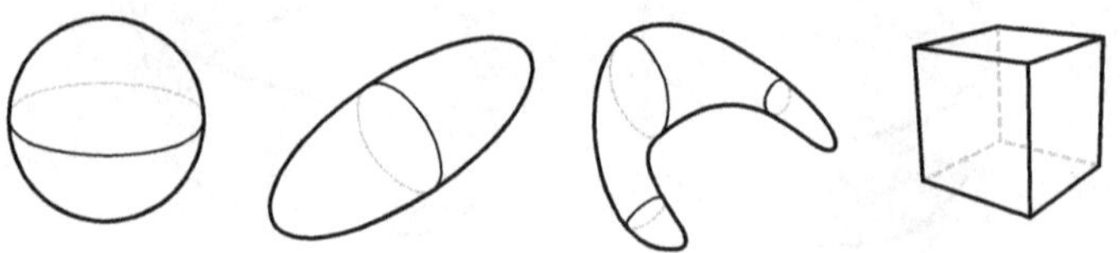

Closed surfaces that are the "same" as a sphere.

"Oh, neat!" Tetra said.

"So let's look at a closed surface that's 'different' from a sphere. For example, a torus."

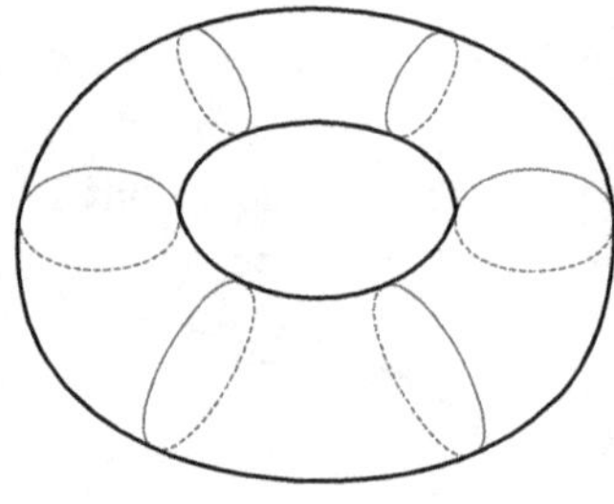

A torus.

"A donut?" Tetra said.

"The skin of a donut," Miruka replied. "Like with the sphere, there's nothing inside. Despite that similarity, however, a sphere and a torus are different closed surfaces."

"Because no matter how you deform a sphere, you can't make it into a torus, right?" I said.

"That's right," Miruka said. "Of course, we also need a mathematical definition of what it means to 'deform' something. Roughly put, it means we can stretch or shrink the surface like it's made of rubber, but we aren't allowed to cut it, or poke holes in it."

"I see," Tetra said. "So we have spheres and tori and... hmmm, what else? It seems like there should be other kinds of closed surfaces, but I can't think of one."

"Well, a torus kind of looks like a float for one person, right? Imagine what a float for two would look like. A closed surface like that, one with two 'holes' in it, would be different from both a sphere and a torus."

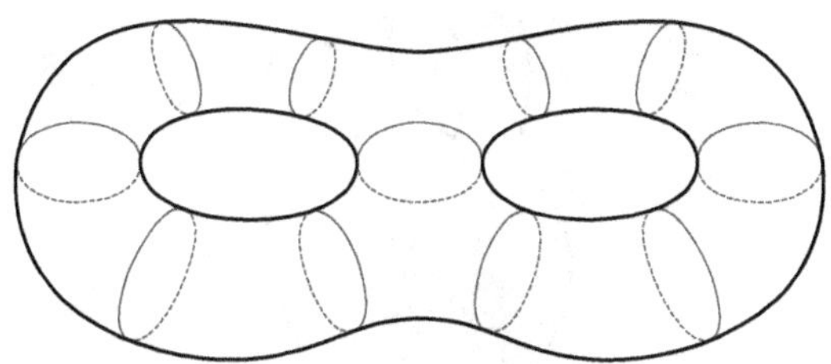

A float for two people.

"And you could keep doing that, right? Just create a float for three people, one for four people, and so on."

"You could," Miruka said, "and every one of them would be different. You can even think of a sphere as a float for zero people. In that sense, the only orientable closed surfaces are floats for n people, meaning we can classify those surfaces according to their number of holes."

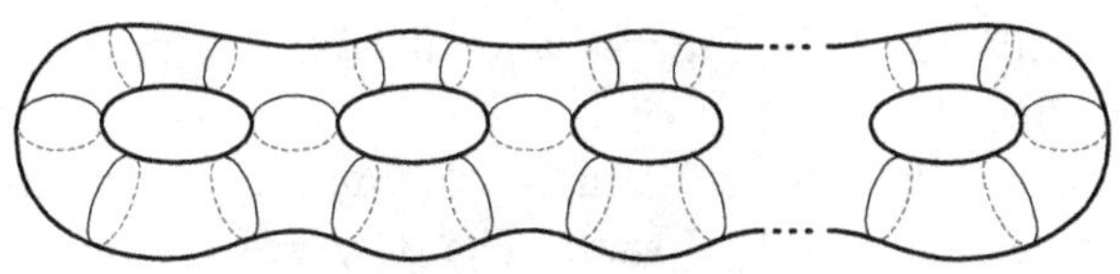

A float for n people.

"Okay," Tetra said. "I'm getting good with orientable closed surfaces, but..."

Miruka raised an eyebrow. "But?"

"But what would a closed surface that *isn't* orientable look like? I guess it would have just one surface, like a Möbius strip, but one with no boundaries? I can't imagine..."

"We need to bump the Möbius strip up to higher dimensions," I said.

"That's right," Miruka said. "A Klein bottle."

2.3.5 Non-orientable Surfaces

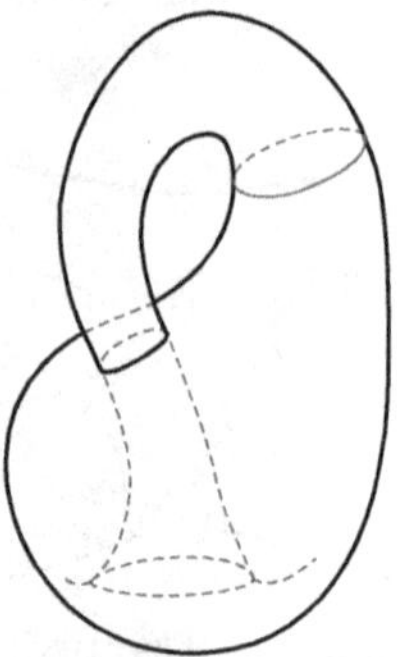

A Klein bottle.

This reminded me of when Miruka and I went to an amusement park, where we had built a Klein bottle out of Lego blocks. And then—

I also wondered if Miruka had been pondering her future back then.

"Hang on!" Tetra nearly shouted. "A closed surface isn't supposed to have boundaries, right? But this Klein bottle is, like, poking right through itself! The hole it makes when it does that makes a boundary, right? Right?"

"That's just because a Klein bottle can't be represented as a three-dimensional object," I said. "The only way to draw one is to make it look like it has a hole in itself that it passes through."

"Indeed," Miruka said. "If you try to represent a Klein bottle in three dimensions, you can't avoid having it intersect itself. Let's just ignore that for now and confirm that we can't distinguish between its front and back surfaces. If you imagine starting to paint one 'side' of a Klein bottle, at some point you find yourself painting its interior, until the entire thing is colored."

"Okay, sure. There's still something about that hole that's bothering me, but otherwise I do see what you're saying. I also see how this is a lot like what happens when you color a Möbius strip. How when you're trying to just color its front, if you can call it that, you find yourself wrapping around to, well, its back."

"They are similar," I said. "That's why I said a Klein bottle is something like a higher-dimension version of a Möbius strip."

"Actually, you can create a Klein bottle by attaching two Möbius strips to each other," Miruka said.

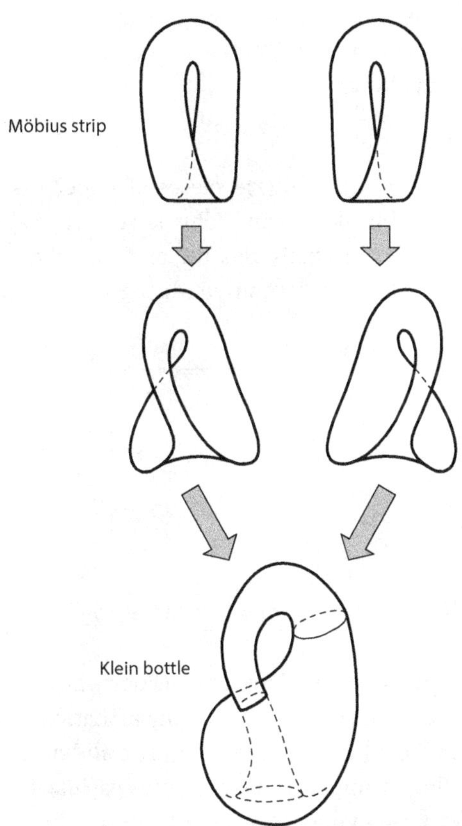

Creating a Klein bottle by attaching two Möbius strips.

"Well look at that!" Tetra said.

"Very cool," I agreed.

Tetra started counting on her fingers. "So we have spheres, and tori, and floats, and Klein bottles, and... Hmmm, any other kinds of closed surfaces?"

I spent a moment trying to create some new form in my head, but didn't come up with anything.

"There is one good tool for looking for them," Miruka said. "A more comprehensive method that doesn't involve directly manipulating the figures we draw. That way, we don't have to worry about self-intersections."

"And what way is that?" Tetra asked.

"Nets, Tetra. Nets."

2.3.6 Polygonal Nets

"Pretend you have some square pieces of paper, like what you use to fold origami," Miruka began. "This is very special paper, though. It's made from an extremely flexible material that you can stretch as much as you like. Also, when you bring two edges together, they stick."

Extremely flexible, stretchable origami paper.

"When we join edges we have to consider whether there's twisting involved, and that means we need to pay attention to the orientation of those edges. So when we join edges, let's add arrows to show their orientation. For example, if we add arrows like this, we have the polygonal net for a cylinder."

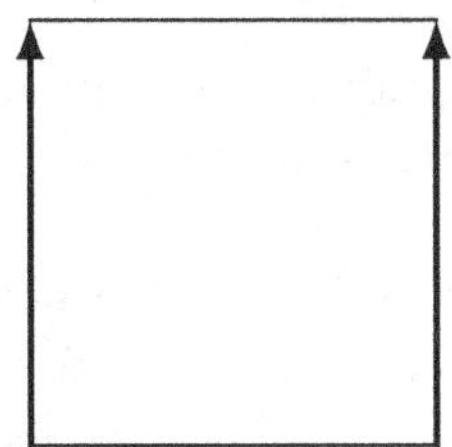

Polygonal net for a cylinder.

"This means we're joining the right and left sides, right?" Tetra said, making a curving gesture with her hands. "Like this."

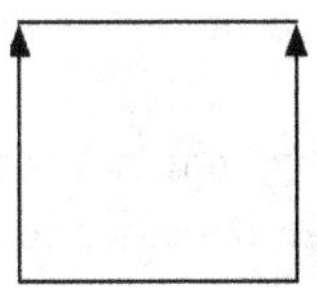
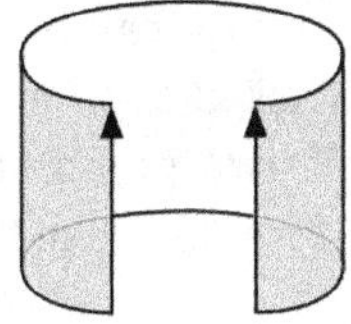
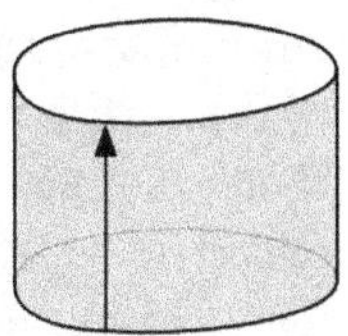

Joining edges to form a cylinder.

Miruka nodded. "But if we flip the direction of the arrows, we get the net for a Möbius strip."

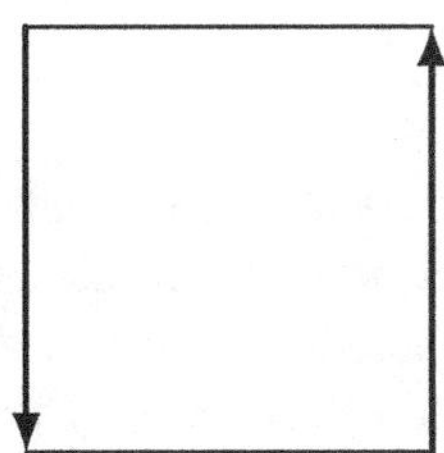

Polygonal net for a Möbius strip.

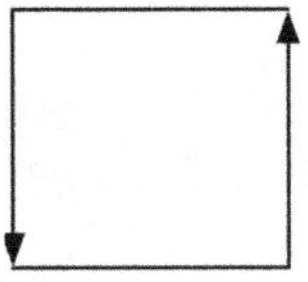
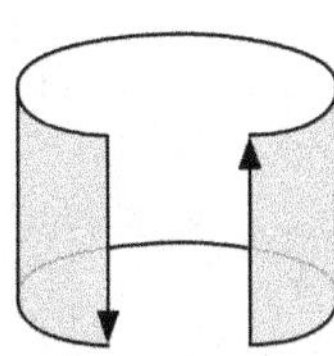
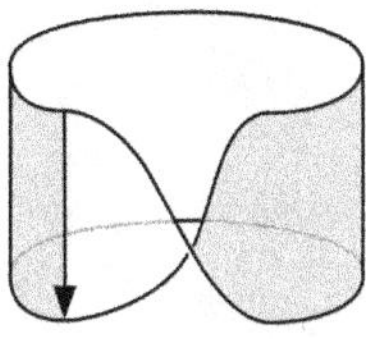

Joining edges to form a Möbius strip.

"Ah, right," I said. "Lining up the arrows makes the half twist. In fact, so long as the arrows line up, you'll have an odd number of twists, no matter how you made them."

"Creating these nets also make the boundaries clear," Miruka added. "Edges with arrows will be joined, so they can't be boundaries, only unlabeled edges can."

"Yeah, sure enough."

"But that's not all," Miruka continued. "So long as we can correspond forms and nets in our heads, there's no need to actually join the edges in three dimensions. We just consider pairs of edges with aligned arrows as being the same. When we talk about joining figures in mathematics, we're talking about this kind of sameness."

"Sameness, how?" Tetra asked.

"For example, let's name two points in the polygonal net for a Möbius strip as A and A′. If we do it like this, we can consider those as being the same point."

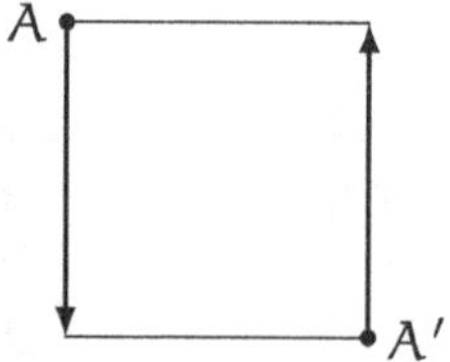

"But . . . they're in different places, right?"

"They only look that way in the diagram. Think about what happens when we join the edges. See how A and A′ end up in the same place? And we can do that for every point on the joined edges, so we can consider the entirety of those edges as being the same."

"We can do that for the endpoints too," I said, "naming them B and B′. Joining makes them the same point."

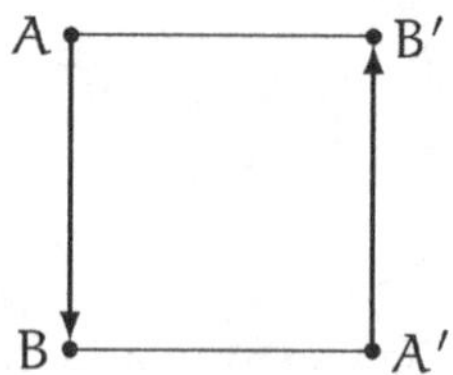

"Okay, I see."

"Moving on, after we've joined the edges with arrows, boundaries remain. What we want to do is create a closed surface, in other words a surface that has no boundaries. That means we have two boundaries to remove. How do we do that?"

"We... join those too?" Tetra said.

"That's right, Tetra." Miruka pointed a finger at her. "So do it."

Problem 2-1

What closed surface does this polygonal net describe?

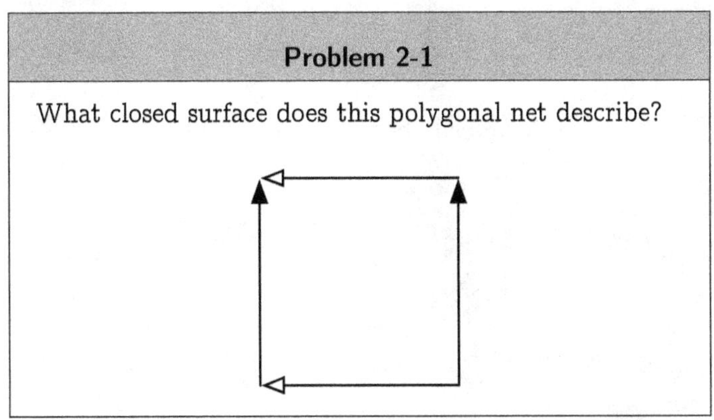

"There are no edges without arrows here," Miruka said. "Also, there are two each of two kinds of arrows. If we join same-type arrows, we get a shape with no borders, in other words a closed surface. The question is, what kind?"

"A sphere, maybe?" I suggested. "We're connecting right to left, top to bottom, right?"

"What do you think, Tetra?"

"I think it's... a torus? Maybe?"

"A torus?" I said. "Wait, you're right! How'd you get that so fast?"

"Well, you see this in computer games, right? Like, when a ball goes off the right side of the screen and reappears on the left, or goes up and reappears from the bottom. A screen like that is the surface of a torus."

"Very good," Miruka said. "When you connect the tops and bottoms you get a cylinder, which has two boundaries. When you join those boundaries, paying attention to their orientation, you get a torus."

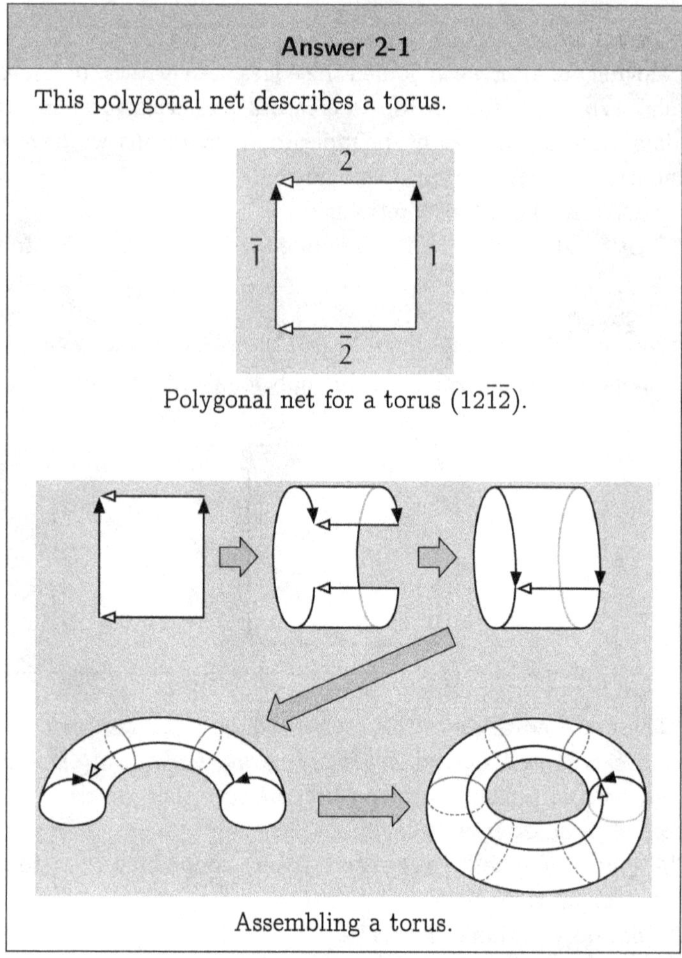

Answer 2-1

This polygonal net describes a torus.

Polygonal net for a torus ($12\bar{1}\bar{2}$).

Assembling a torus.

"What's this $12\bar{1}\bar{2}$ mean?" Tetra asked. "Edge numbers or something?"

"It's a way to represent a polygonal net as a sequence of numbers. We assign the same number to edges that we're joining—in other words, edges we're considering to be the same. Imagine we're following the edges of our origami paper counterclockwise. If the direction we're moving in coincides with the direction of an arrow, we just leave the number as-is. But if the arrow points in the opposite direction, we put a bar above the edge's number. So $12\bar{1}\bar{2}$ describes the net for a torus."

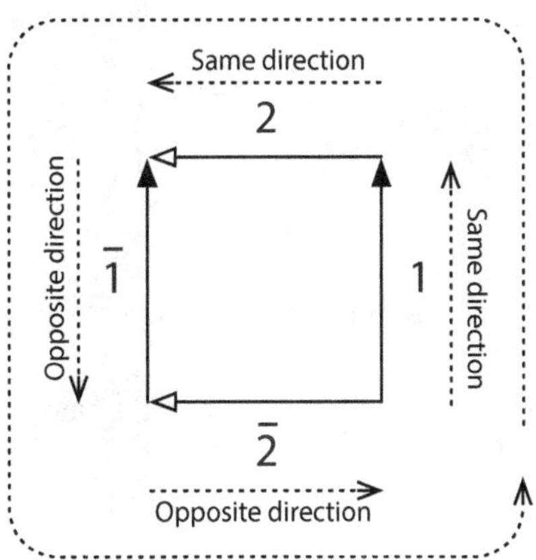

"Our next example will answer your question," Miruka said, looking straight at me.

"My question?" I said.

"Well, at first you thought this first example was a sphere, right? So you must be wondering just how we would use stretchy squares to create a sphere. The answer is, like this."

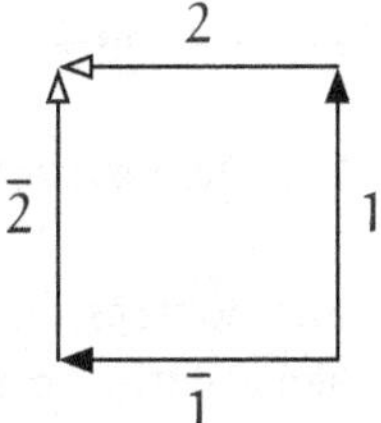

Polygonal net for a sphere $(12\bar{2}\bar{1})$.

"Ah, interesting. The nets for a torus and a sphere are very different, aren't they? $12\bar{1}\bar{2}$ for the torus, $12\bar{2}\bar{1}$ for the sphere."

Tetra clapped her hands together. "It reminds me of making potstickers, the way you squeeze the edges of the dough wrapper together to make them stick! It's like we're joining the edges and inflating it to make a ball!"

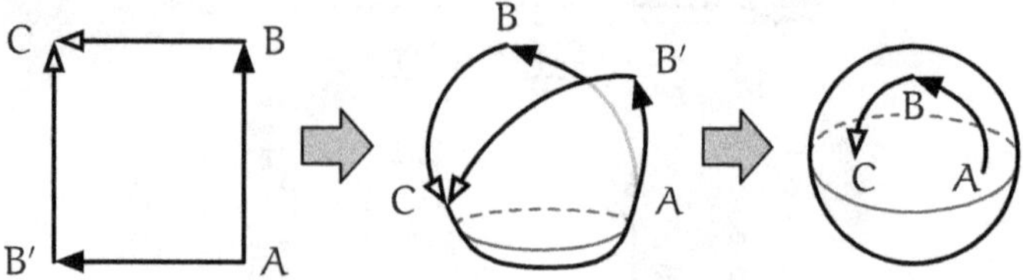

Assembling a sphere.

"If you want to make potstickers," Miruka said, it would be easier to use $1\bar{1}$ instead of $12\bar{2}\bar{1}$. Of course in this case we aren't using square origami paper, we're using sheets with just two sides."

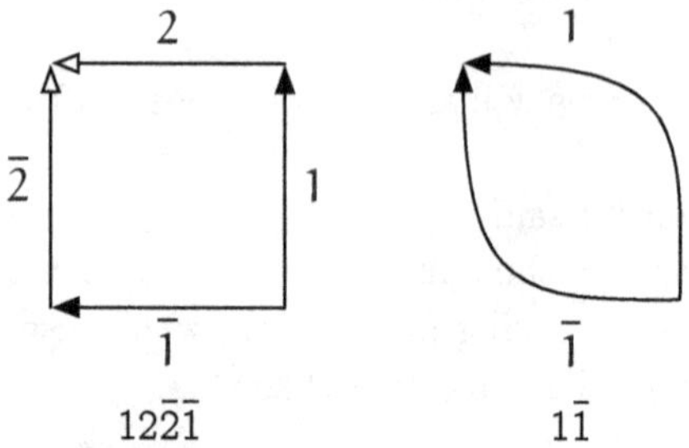

Two polygonal nets for a sphere.

"Oh, so we can make different nets for the same shape! Neat!" Tetra said.

"So we've done toruses and spheres. What's next?"

"I think I see how to make a Klein bottle," I said.

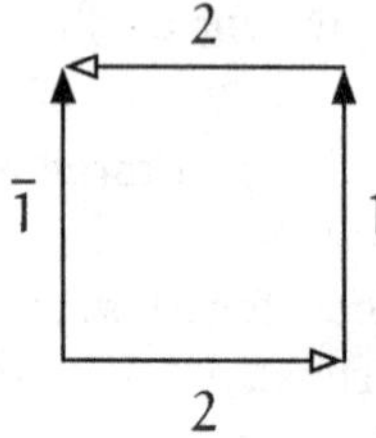

Polygonal net for a Klein bottle ($12\bar{1}2$).

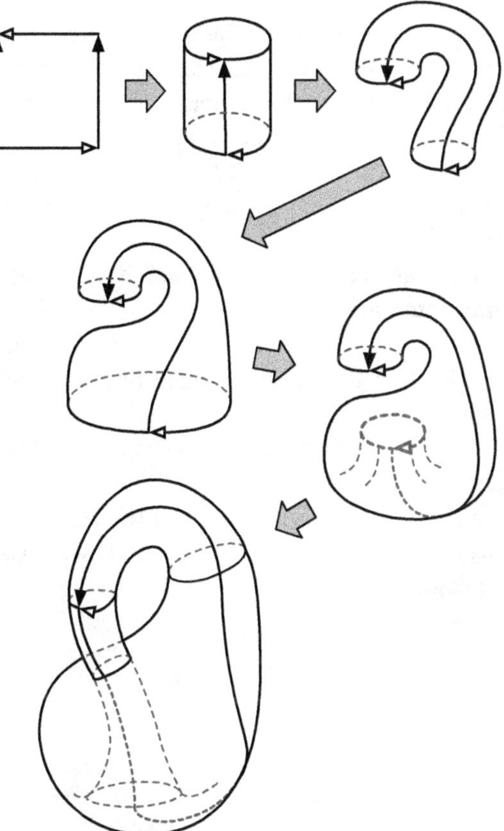

Assembling a Klein bottle.

"Oh! So a Klein bottle is like a twisted torus!" Tetra said, folding her arms into a shape like a tangled snake.

"So just how many ways are there to join opposing edges of a square, I wonder? If we join two opposing edges with the same orientation we get a cylinder, and if they have opposite orientations we get a Möbius strip. Then there are two ways to join the two edges that are remaining in the map, one for each orientation—starting with the cylinder, if the orientations are the same we get a torus, and if they're opposite we get a Klein bottle. But ..."

"We can also join the two remaining edges in the net for the Möbius strip, right?" Tetra said.

"Yeah, but I think we only get Klein bottles from a Möbius strip. Because joining the two remaining edges in the cylinder net with opposite orientations is the same as joining the two remaining edges in the Möbius strip net with same orientations."

"Oh, right, okay. But that's just for the same orientations in the Möbius strip net, right? Can't we put those two remaining edges in opposite orientations?"

"I don't think we can do that, can we? It seems like something would run into something else."

"And what's wrong with that?" Miruka asked. "You didn't have any problem with the Klein bottle intersecting itself. Why can't we do that again?"

"I guess, but ..." I grimaced. "What shape would we get? I just can't imagine what that would look like."

"When we give the remaining edges in the Möbius strip net opposing orientations and join them, we get a closed surface like this. It's called a projective plane."

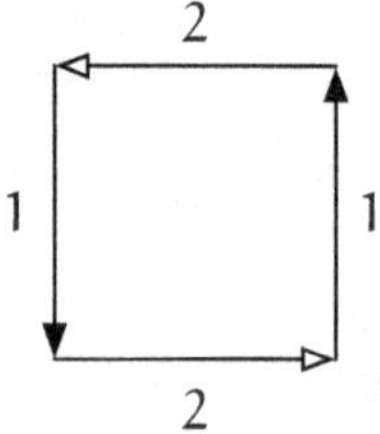

Polygonal net for a projective plane (1212).

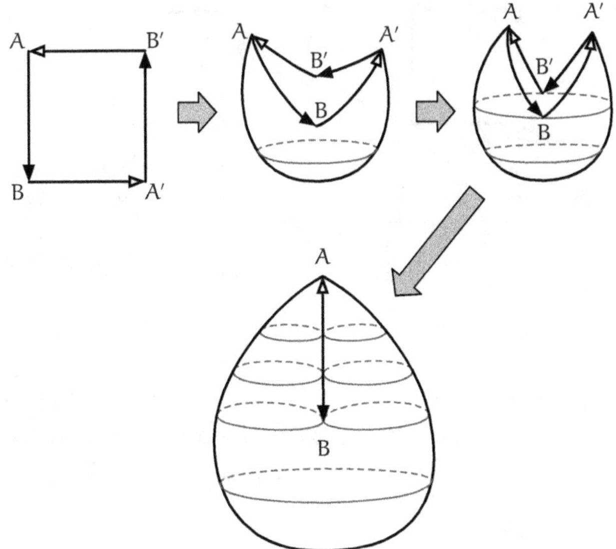

Assembling a projective plane.

"Well that's... interesting," I said. "Let's see, we're joining the A-to-B arrow with the A′-to-B′ arrow, and we're simultaneously joining the B′-to-A arrow with the B-to-A′ arrow."

"Hmmm, let me think," Tetra said. She twisted her hands around while staring at the polygonal nets until finally nodding in satisfaction. "Okay, that's all of them."

"Looks like it. So if we join the cylinder boundaries as-is we get a torus, and if we give it a half-twist we get a Klein bottle. We also get a Klein bottle if we join the boundaries of a Möbius strip as-is, and if we give those a half twist we get a projective plane. Okay, that's all the ways we can join opposing edges."

"A summary!" Tetra said as she started sketching nets.

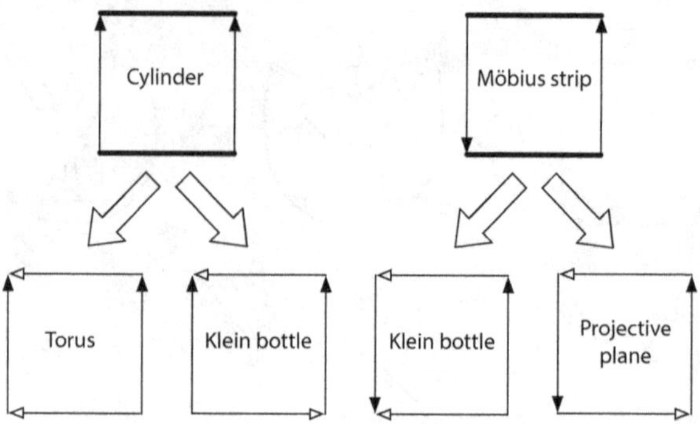

"Just like with the sphere, we can also create the projective plane with a two-sided shape," Miruka said. "With the sphere we changed $12\bar{2}\bar{1}$ to $1\bar{1}$, and to make a projective plane we can change 1212 to 11."

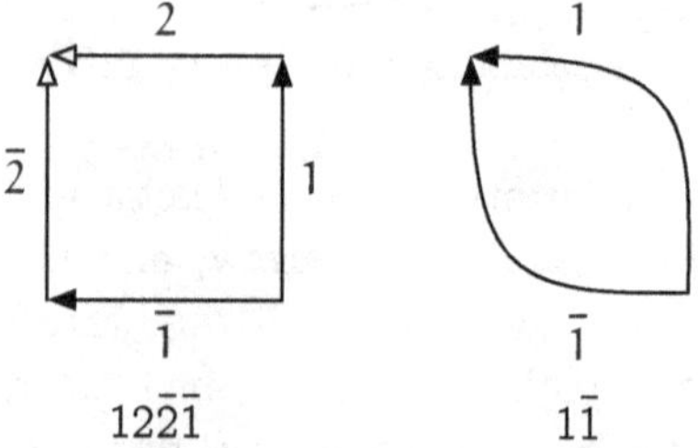

Two polygonal nets for a sphere.

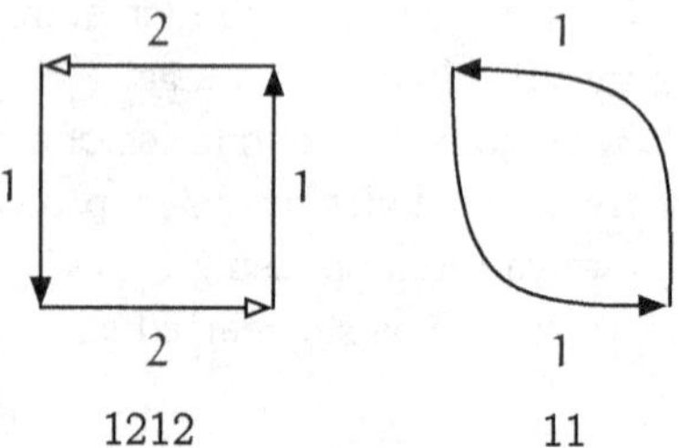

Two polygonal nets for a projective plane.

"So you're, like, combining the arrows?" Tetra said, peering at the nets.

"Interesting," I said.

"Hey, that reminds me! I guess we've covered the torus, but we haven't made the two-person float."

"That's because we were mainly working with squares," Miruka said. "We *can* make the two-person float, but we need to use octagonal origami paper instead."

"So we're going to join edges on an octagon? Wow, that's hard to imagine."

"It's also getting a bit ahead of ourselves. Before that, we should talk about connected sums."

2.3.7 Connected Sums

"What's a connected sum?" Tetra asked.

Miruka cocked her head. "Think of it as... cutting a small disk in two shapes, then connecting their edges to make a new shape. Here's how we do that for two tori."

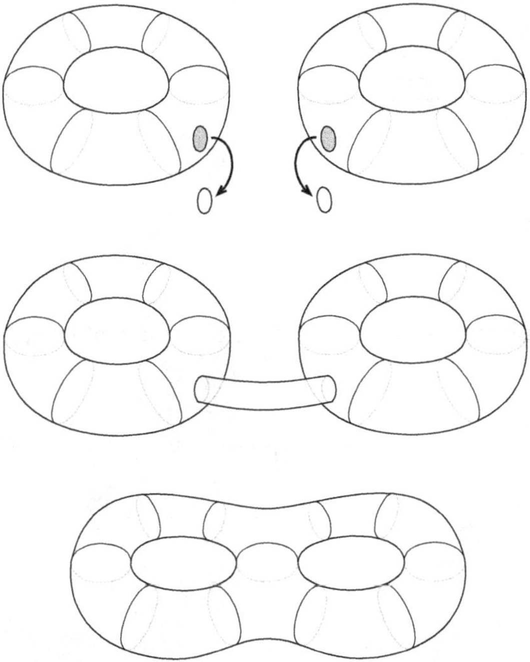

Creating the two-person float shape by a connected sum of two tori.

"See? When we consider the boundaries on the disks we cut out as being the same, we get the two-person float shape."

"I guess," I said. "But polygonal nets with all this cutting and joining holes somehow seems even worse than octagons."

"Not at all," Miruka said. "It's quite easy to create a net for the two-seater float. First, place two torus nets next to each other."

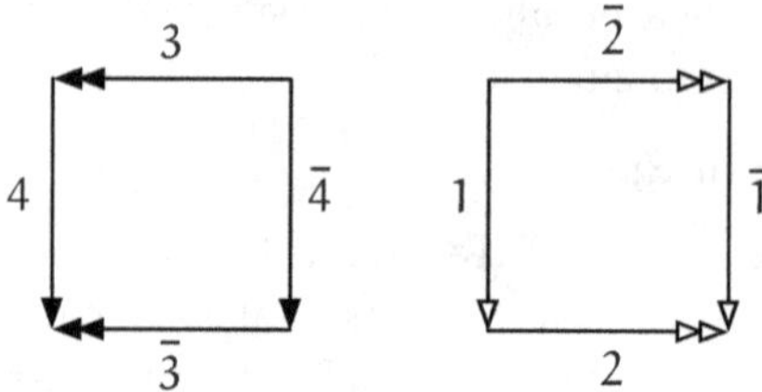

"Next we cut disks from both tori. We're considering their boundaries to be the same, so we'll name them 0 and $\bar{0}$."

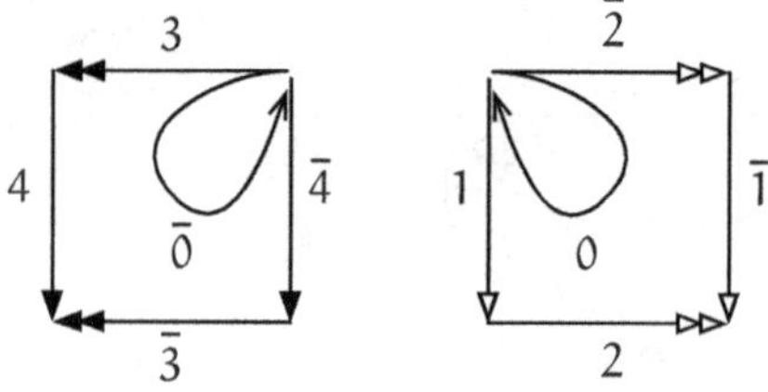

"Now we stretch these edges to make two pentagons."

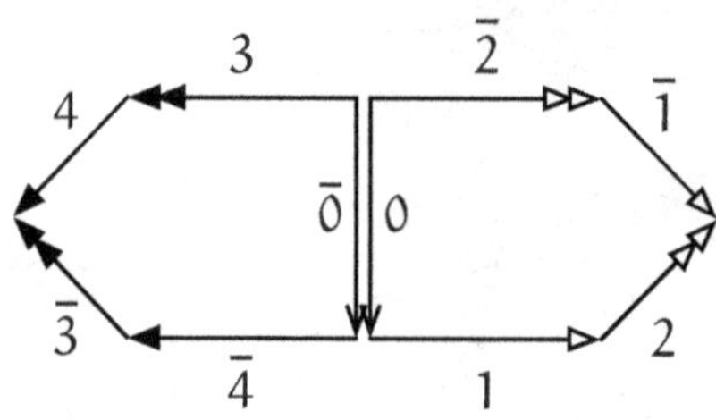

"If we join 0 and $\bar{0}$, we get an octagon."

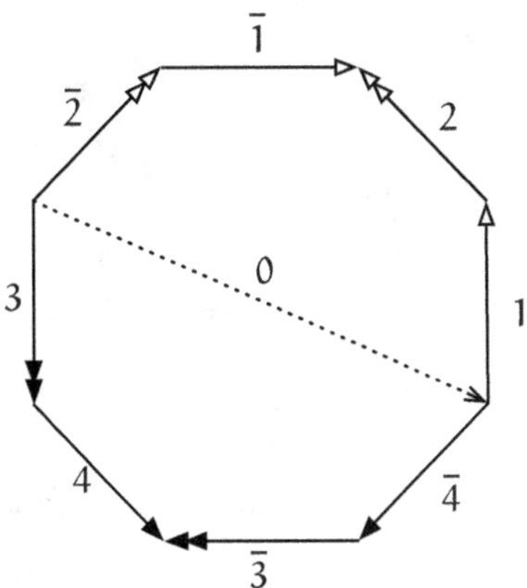

Polygonal net for the two-person float shape ($12\overline{1}\overline{2}34\overline{3}\overline{4}$)

"This is a closed surface of genus 2, what we've been calling the two-person float shape."

"So it is possible!" Tetra said.

"I think I see the pattern here," I said. "The sphere was $1\overline{1}$, the torus was $12\overline{1}\overline{2}$, the float was $12\overline{1}\overline{2}34\overline{3}\overline{4}$, so—"

"Very good. We can categorize orientable closed curves like this."

- $1\overline{1}$ is a sphere (closed surface with genus 0).
- $12\overline{1}\overline{2}$ is a float for one person (closed surface with genus 1).
- $12\overline{1}\overline{2}34\overline{3}\overline{4}$ is a float for two people (closed surface with genus 2).
- $12\overline{1}\overline{2}34\overline{3}\overline{4} \cdots (2n-1)(2n)(\overline{2n-1})(\overline{2n})$ is a float for n people (closed surface with genus n)

"We can also think of a float for n people as the connected sum of a sphere and n tori. Cut n disks out of the sphere—"

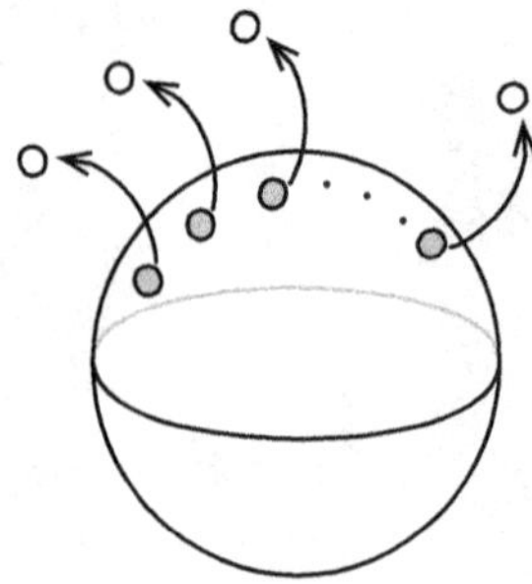

"—then attach n tori with disks removed to the cutouts on the sphere."

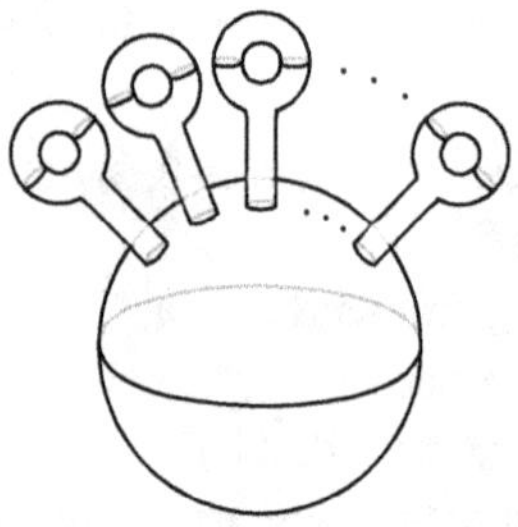

"And there you go—a float for n people."

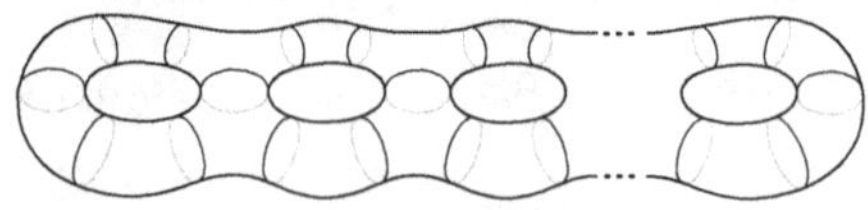

"Non-orientable closed surfaces, on the other hand, are projective planes, Klein bottles, or connected sums of spheres and n projective planes."

- 11 is a projective plane.
- 1122 is a Klein bottle.
- $1122 \cdots nn$ is a connected sum of a sphere and n projective planes ($n = 3, 4, 5, \ldots$).

"By the way, we can also think of a projective plane as a connected sum of a sphere and one projective plane, and we can consider a Klein bottle as a connected sum of a sphere and two projective planes. So just the phrase 'connected sum of a sphere and n projective planes' is enough to cover all non-orientable closed surfaces."

- $1122 \cdots nn$ is a connected sum of a sphere and n projective planes ($n = 1, 2, 3, \ldots$).

Categorization of closed surfaces (connected sums)

- Orientable: A connected sum of a sphere and n tori ($n = 0, 1, 2, \ldots$)
- Non-orientable: A connected sum of a sphere and n projective planes ($n = 1, 2, 3, \ldots$)

A connected sum of a sphere and n tori.

A connected sum of a sphere and n projective planes.

"Okay, hang on. Let me back up a bit," Tetra said. "A Klein bottle is 1122? I thought it was $12\bar{1}2$?"

"It can be either one," Miruka said.

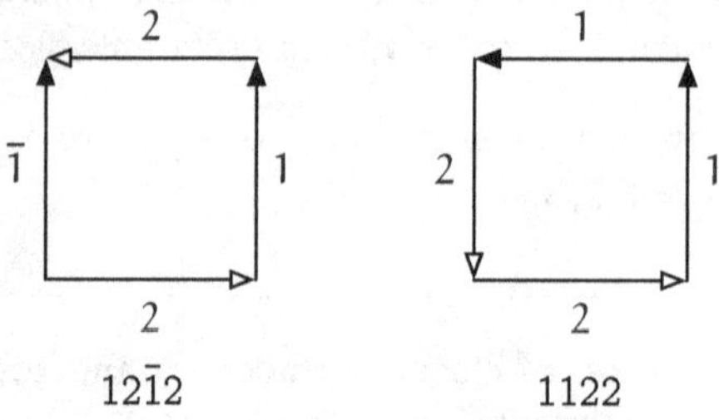

Polygonal nets for a Klein bottle.

"Wait, how's that?"

"Just cut along a diagonal and renumber."

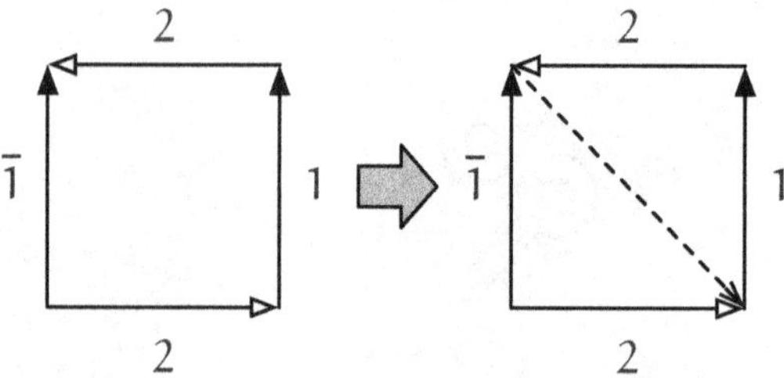

Cut along the diagonal on the polygonal net for a Klein bottle ($12\bar{1}2$).

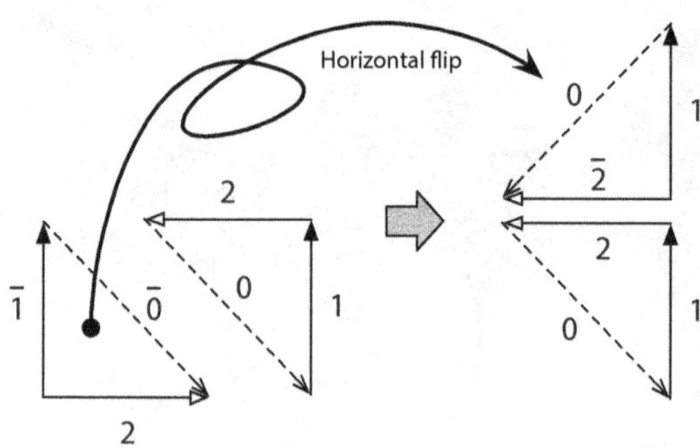

Flip one side and join to 2.

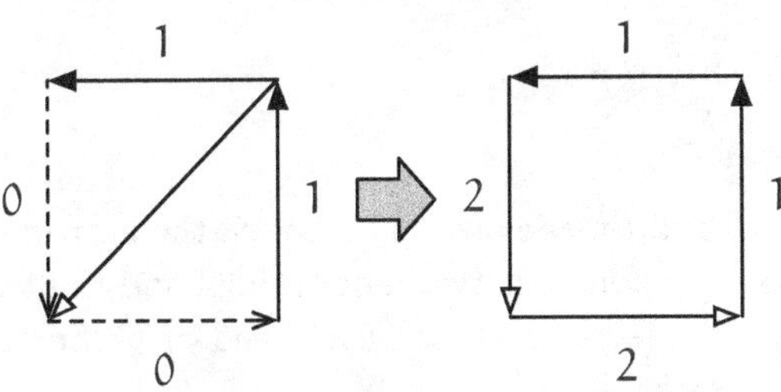

Renumber to make a new polygonal net (1122).

"A torus with a disk removed from it is called a handle," Miruka continued. "A projective plane with a disk removed is a Möbius strip. So we can also categorize closed surfaces like this."

Categorization of closed surfaces (connected sums)

- Orientable: A sphere with n disks removed and n handles attached ($n = 0, 1, 2, \ldots$)
- Non-orientable: A sphere with n disks removed and n Möbius strips attached ($n = 1, 2, 3, \ldots$)

Tetra held her head in her hands. "We remove a disk from a projective plane and get ... a Möbius strip?"

"And we're attaching Möbius strips to holes?" I said. "We can do that? Again, hard to imagine ..."

"Yes, we can do that," Miruka said. "The boundary of a disk is a single closed curve. The boundary of a Möbius strip is also a single closed curve. Need I say more?"

"The library is *closed*!"

We all jumped at the sudden proclamation. Looking behind us, we saw Ms. Mizutani, the school librarian, in her customary dark glasses and tight skirt. She spent most of her time in her office, emerging only to stand in the middle of the library to announce its closure. I hadn't realized how late it had gotten.

We hurried to collect the many pieces of paper scattered about, and headed out.

2.4 Heading Home

2.4.1 *Like a Prime*

Miruka, Tetra, and I made our way through the narrow streets that wound among the houses between our school and the train station. I walked next to Tetra, with Miruka ahead of us both. Tetra was uncharacteristically silent.

"You okay?" I asked.

"I was just thinking about how a Möbius strip is kinda like a prime."

Miruka turned to face us, a quizzical look on her face. "A prime, you say?"

"Um, right. You know how we can perform prime factorization for any integer? And how that means we can create any integer as a product of primes? Well, when you were talking about categorizing closed surfaces, it reminded me of that. Maybe that's not the right way to think of it, but still..." Tetra continued more slowly, carefully choosing her words. "Mmm, there's still a lot about this I don't understand, but if we can create any closed surface from handles, Möbius strips, and spheres with holes punched into them, doesn't that make the Möbius strip a really important part? Something like a prime?"

"Hmph." Miruka nodded.

"Still, it takes some serious mental gymnastics to imagine using a Möbius strip to plug a hole," I said. "Cutting out disks to create connected sums in polygonal nets, using nets to visualize Klein bottles... It's all like working on puzzles. Using infinitely stretchable origami paper to make shapes? Topology is... interesting."

> Can we list all possible two-dimensional manifolds? After Magellan's expedition returned, but before anyone had explored the poles, could we have been ready with a set of possible shapes for our world? The answer, together with its proof, is one of the great achievements of the nineteenth century.
>
> Donal O'Shea
> *The Poincare Conjecture* [4]

Chapter 3

Near Tetra

> But actually time isn't a straight line. It doesn't have a shape. In all senses of the term, it doesn't have any form. But since we can't picture something without form in our minds, for the sake of convenience we understand it as a straight line.
>
> Haruki Murakami
> *1Q84*

3.1 Near My Family

3.1.1 Yuri

On my way to the train station the next morning, I ran into Yuri.

"Well look who it is," I said.

"Howdy, cuz! Going my way?"

Despite our living so close together, I didn't often run into Yuri during my morning commute. She was always wearing jeans when she came to my house, so seeing her in her school uniform was also a rare occasion.

"What are you staring at?" she said.

"Nothing, nothing."

"Oh, how's your mom? Is she in the hospital?"

"No. Well, sorta. But she just stayed for some tests, not treatment. Apparently there's nothing to worry about, but she's using this as an opportunity to get a comprehensive checkup."

"Well that's good to hear."

"So anyway, no need to go visit her in the hospital or anything. Be sure to tell your mom that, too."

"I will. She probably already knows, though."

"Yeah, word travels fast. Speaking of which, why are you out so early this morning?"

"I'm trying to get to school earlier lately." Yuri peered up at my face. "By the way, it's cool if I come to your house this weekend, right?"

"Since when do you actually ask?"

"I mean, with your mother and all. If she isn't feeling well, I don't want to be a bother."

"Nah, no bother. She should be home and feeling fine by then."

"Great. Well then, tell her I'm looking forward to seeing her."

"I'm sure she'll appreciate that."

Home had felt kind of tense for the few days my mother had been away. My father, who usually didn't leave work until late at night, was there in the evenings. We split the housework my mother normally did between us. It was an interesting illustration of the tenuousness of the daily routines we take for granted. In truth, our daily lives are suspended in a delicate balance, and through living together, families learn how to maintain that balance. Yuri was close to being a family member, but by living apart from us she wasn't part of that balancing act.

Such an odd thing, families. We're part of one from the time we attain awareness, and grow up living in immediate proximity to it. That makes it easy to assume our own family as normal, but of course families come in many forms. No matter how they're made up, however, we consider them to be essentially the same. So what, exactly, *is* a family?

Yuri gave me an odd look. "Something on your mind?"

"No, no. I was just thinking about the form of families."

"Ah. Another weird math thing, of course."

3.2 Near Zero

3.2.1 Practice Problems

I spent that day's study hall working on math, doing practice test questions from one of my exam prep books. I decided I'd use sheets of paper with plenty of whitespace and set time limits to simulate what I'd face in my actual entrance exams.

I had noticed some time back that answer sheets for those tests don't have lines like notebook paper, and since then I'd studied math using unlined paper. There's something very different about doing math on blank paper, as opposed to the lined notebooks I was used to. You have to plan out an easy-to-read answer considering the amount of space you're allocated, and be careful to keep everything straight.

I put my watch on my desk, and turned to the first practice problem.

Problem 3-1

Give an example of a function $f(x)$ defined for all real numbers for which $\lim_{x \to 0} f(x)$ exists and is finite, and $\lim_{x \to 0} f(x) \neq f(0)$.

Interesting.

$\lim_{x \to 0} f(x) \neq f(0)$ implies that

- the limit of $f(x)$ when $x \to 0$ and
- the value of $f(x)$ when $x = 0$

would be different.

No problem. I just had to think of a function with a gap at $x = 0$. So, something like this?

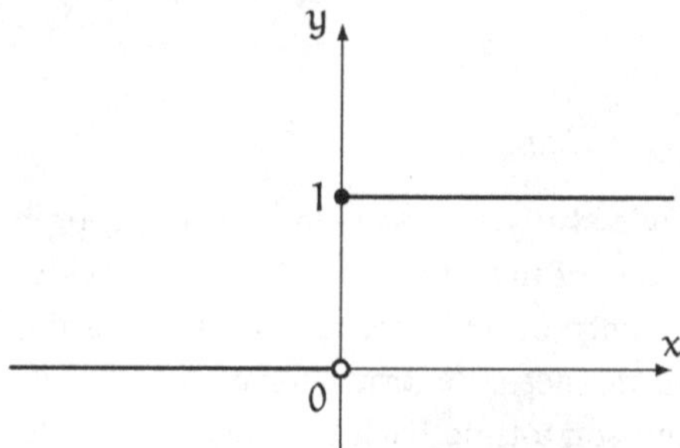

In other words, I could define $f(x)$ like this to provide the kind of example the problem was after.

$$f(x) = \begin{cases} 0 & \text{if } x < 0 \\ 1 & \text{if } x \geqslant 0 \end{cases}$$

Instakill! What's next?

—Wait, something's not right here. I need to read that one more time...

Problem 3-1 (once again)

Give an example of a function $f(x)$ defined for all real numbers for which $\lim_{x \to 0} f(x)$ exists and is finite, and $\lim_{x \to 0} f(x) \neq f(0)$.

This problem puts the following conditions on $f(x)$:

- The function $f(x)$ is defined for all real numbers.
- The limit $\lim_{x \to 0} f(x)$ exists and is finite.
- $\lim_{x \to 0} f(x) \neq f(0)$.

I needed to be sure I had covered each of these. First off, my $f(x)$ was defined for all real numbers, and...

Oops.

$\lim_{x \to 0} f(x)$ doesn't exist! For it to exist, the function would have to approach the same value no matter how x closed in on 0. With my function, when x approached 0 from the positive side $f(x)$ approached 1, but from the negative side $f(x)$ approached 0.

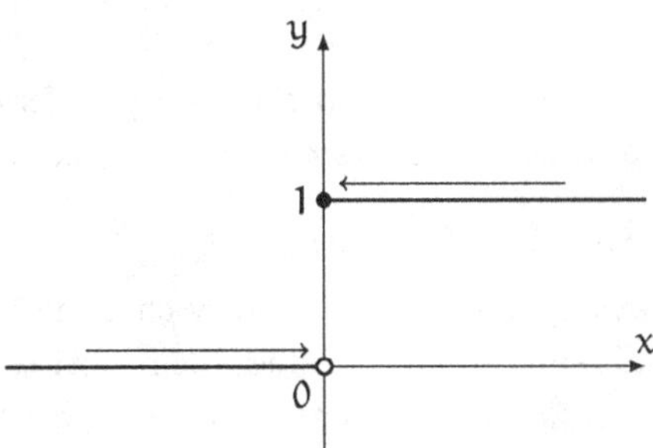

Okay, so this $f(x)$ didn't answer the question. I could fix that, though, by defining a function that takes a different value only when $x = 0$. Something like this.

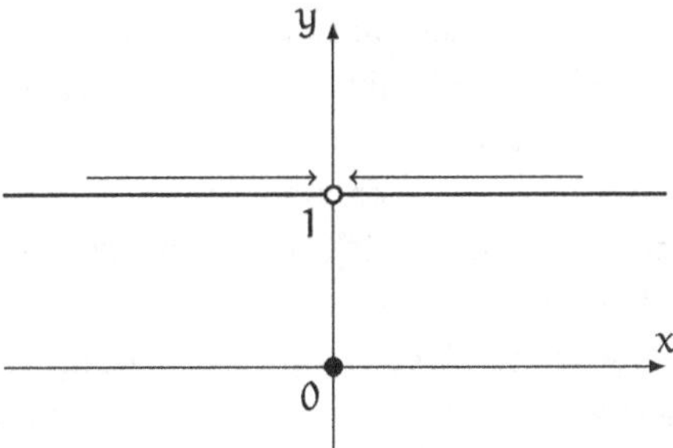

In this case, the limit of $f(x)$ when $x \to 0$ would be 1, but when $x = 0$ the value of $f(x)$ would be 0. So, $\lim_{x \to 0} f(x) \neq f(0)$.

Answer 3-1 (one possibility)

$$f(x) = \begin{cases} 0 & \text{for } x = 0 \\ 1 & \text{for } x \neq 0 \end{cases}$$

I was a little bit worried about how long I'd spent on such a trivial problem. I understood limits, and if I hadn't rushed on this problem there wouldn't have been anything hard about it. But the time limits inherent to tests made me want to rush. That was one reason I feared tests.

No, no... Let's not go there. This is why I practice. Just clear your head and move on to the next problem. Take a deep breath and keep at it, or you're just wasting your time.

3.2.2 Congruence and Similarity

After school that day, I was in the library chatting with Tetra.

"I've been thinking about tori and Klein bottles ever since the other day," she said. "I'm starting to understand how in topology we're able to, like, softly deform shapes. I think. But maybe not."

I sat back to listen to her concerns. She often began these talks with apologies for taking up my time, but once she got started she was hard to stop. Clearly, our topology discussion with Miruka the other day was still on her mind.

"I think what's bothering me is that we haven't seen any equations," she continued. "It feels like we aren't really doing math when we just draw some pictures and that's it."

"You're starting to sound like me," I said.

"In a different sense, maybe. I just get worried that without equations, we don't have any way to make sure we aren't making mistakes. We just say we're stretching here, joining there... Is that really okay, in a math sense? How do we avoid errors?"

"Yeah, you definitely sound like me."

"No teasing, now." Tetra pantomimed a punch in my direction.

I turned serious. "In those polygonal nets Miruka showed us, she used numbers like $11\bar{2}2$ to describe closed surfaces, right? That's probably what we should be looking at. They allow us to make comprehensive investigations, because we can use them to think of changes to polygonal nets as changes to numbers instead, so I wonder if we can't use those in a similar way as equations?"

"I also hadn't really thought much about what it means for shapes to be the 'same,' so I really didn't understand that at all." I recalled when I'd first met Tetra. She'd been worried about how deeply she understood things even then. I remembered the day of our first long talk, in a tiered auditorium here at our school. I couldn't quite remember what we'd talked about, though. "From what I've seen, topology groups together shapes that we would normally say are completely different. I mean, even triangles and squares and circles would be categorized as being the same!"

"Sounds like you're still thinking in terms of congruence and similarity."

"I . . . I am?"

"Yeah. Remember that if two shapes are congruent, they're the same in terms of both size and form. Two of the same shape aren't congruent if their sizes are different."

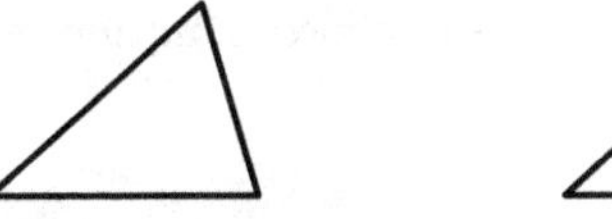

Congruence (same shape and size).

"Sure, I remember that."

"For two shapes to be similar, though, they can be different sizes, so long as they're the same shape. Learning about similarity in school was the first time I ever thought of form and size as separate things. I'd never looked at two shapes and thought, 'They're the same shape, but at different scales.' "

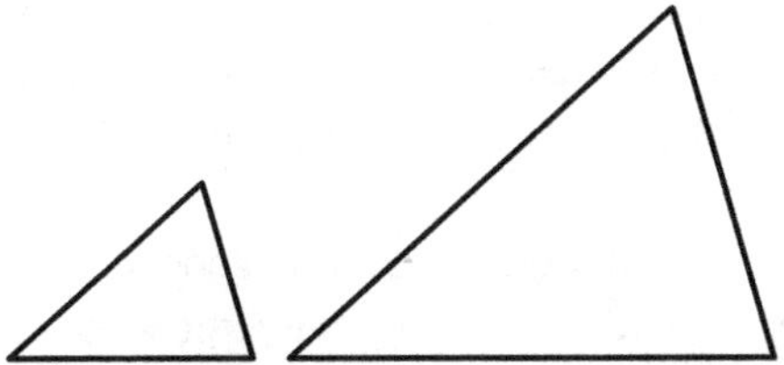

Similar but not congruent (same shape, different size).

"But hang on," Tetra said. "Don't we need to be clear about what we're calling the 'same shape' here? Seems like if we aren't, saying things like 'these shapes are similar' is kind of vague."

"You're right. In the case of triangles, we can say they're congruent if the lengths of their corresponding sides are equal. And we can say they're similar if they have the same ratios of the lengths of corresponding sides."

"Mmm..." Tetra groaned. "So in that case, isn't congruence a special case of similarity? Because similarity is equality of the ratios of corresponding sides, and congruence is the case where that ratio is furthermore $1:1$! Or... did I just say something obvious?"

"Not at all. It's important to go back and confirm your understanding like that. Sure, you can think of congruence as a special case of similarity, one where the ratios are $1:1$. Conversely, you can think of similarity as a generalization of congruence. Just say that congruence is when the ratios are $1:1$, but similarity is when the ratios are $1:r$. Then we've generalized congruence through introduction of a variable."

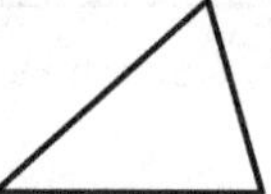

Congruence is similarity with $1:1$ ratios of side lengths.

"Hey, I just thought of something," Tetra said. "Since triangles and circles are the 'same shape' in the world of topology, doesn't it seem like topology too should have words for different kinds of sameness, words like congruence and similarity?"

I looked at Tetra with widened eyes. She was spot on. I had read about topics from topology in many books. I'd seen discussions of why donuts are like coffee cups many times. But I'd never really followed Tetra's line of thought myself.

Tetra had a love of words. She was good at foreign languages, and was always careful about how she expressed herself. She took the time to choose her words, and by being careful to use precise language when talking about math, she was able to maintain a good grasp on what remained unknown to her.

"Still with me?" Tetra asked.

"Just sitting here thinking about what you've become."

"I'm just Tetra," she said, looking straight at me. "Same as always."

3.2.3 Creating Correspondences

That night, my mother was still in the hospital and my father was there visiting her. I went into our kitchen and made myself a cup of coffee. Normally my mother would have headed me off and offered to make me a cup of cocoa instead, but not tonight.

Because she's in the hospital.

And I'm at home, alone.

Every family changes, little by little over the course of time. Change is inevitable. Eventually I would move out to live on my own, changing the form of my family myself. Other than that I didn't know how my family would change, just that it would.

I went back to my room to work on a few problems while drinking my coffee.

You can't read about topology without hearing that hoary joke how topologists are people who can't tell the difference between their coffee cups and their donuts. Coffee cups have a hole in their handle, corresponding to the hole in a donut.

Speaking of correspondences...

I recalled my conversation with Tetra that day. I had told her that congruent triangles were ones having equal corresponding side lengths. This condition for the congruence of triangles had deep roots in my mind, but talking about it today raised a question: what exactly are corresponding sides?

I imagined two congruent triangles, and thought about what their corresponding sides would be. Nothing really mysterious about it, but it seemed like I was just determining which sides correspond by eyeballing them. Something about that didn't seem quite right. After a time, I wasn't even sure what I was thinking about.

So what do I have to do to mathematically establish a correspondence? That's what's hanging me up...

I decided to start from the beginning and work toward my goal.

A figure is a collection of points. That suggests that creating a correspondence between two shapes would also imply a correspondence between their points. Given two shapes, it should be possible to take any point on one and create a unique correspondence with some point on the other. A mapping, in other words. So maybe thinking mathematically about how to correspond shapes with each other is a matter of thinking about mapping one to the other? *Yeah, that sounds right.*

So to consider coffee cups and donuts as being essentially the same shape, I would need to find some clever way of corresponding cup points with donut points. Some slick mapping. If that's the case, there must be some useful definition of mapping in the topological toolbox. Also, that definition should be somehow similar to the mappings for congruence and similarity I was already familiar with.

I'd first met Miruka when I started at my high school. Thanks to the conversations we'd had since then, my understanding of mathematics had become far deeper, including what I knew about sets and mappings. Of course, I'd also learned a lot from reading books, but Miruka's "lectures" had delivered many epiphanies.

I was very lucky to have met Miruka in high school. But now that time was coming to an end. And after graduation...

No, don't think about that. Gotta stay focused on college entrance exams.

It was still fall. Soon I'd have finals. Then it would be winter, and just before Christmas I'd have a practice test simulating an actual exam.

Teachers at my school were always telling us how entrance exams would determine our futures. Well no kidding. It annoyed me just to hear the words. The importance of these tests was a heavy burden. The fact that they were something like a mold that would cast the form of my future made them even heavier.

I spent the rest of the evening thinking such pointless thoughts.

3.3 Near a

3.3.1 Congruence, Similarity, and Homeomorphism

Going to the school library after classes the next day, I found Tetra and Miruka sitting across from each other at a table, talking. I couldn't hear what they were saying, but Tetra's exaggerated body language gave me a good idea. From the way she was moving her hands, it could only have been congruence of shapes. I walked up to their table.

"—so I figured there must be some other word for 'same,'" Tetra was saying.

"Homeomorphism," Miruka said. "The word you're looking for to describe the concepts of congruence and similarity is homeomorphism. Homeomorphic, if you want the adjective."

"Ah, interesting!" Tetra said, entering linguist mode. "'Homeo-' is a prefix meaning 'same,' right? And I'll bet 'morph' means 'form.' Like metamorphosis from a caterpillar into a butterfly! Then '-ism' to make it a noun! Okay, homeomorphism means 'same shape.' Got it!"

"Glad to hear it. Anyway, just like congruence and similarity are fundamental concepts in geometry, homeomorphism is a fundamental concept in topology."

"So you can say, like, this shape and that shape are homeomorphic?"

"That's right. Like coffee cups and donuts."

"I can sort of imagine changing a coffee cup into a donut," Tetra said, moving her hands like a potter sculpting clay, "but I don't see how we can use math to describe such vague concepts."

"I think mappings have something to do with it," I said, sitting next to Tetra. "I thought about this a bit last night, how we can consider congruence and similarity in terms of mappings."

"Well put," Miruka said. "For example, we can think of congruence between two shapes as implying the existence of some mapping that doesn't change the distance between any two points. Also, we can think of similarity between two shapes as a mapping that doesn't change ratios of distances between two points."

"So saying two shapes are homeomorphic is also saying something about the existence of a mapping," I said.

"Exactly. It's that mapping that we call a homeomorphism. So when you say two shapes are homeomorphic, you're saying that a homeomorphism exists between them."

"But hang on, Miruka, those are just names!" Tetra said. "A homeomorphism exists for two shapes that are homeomorphic? That doesn't really tell us much, does it."

"You're right. The problem is that we haven't yet defined just what a homeomorphism is."

"As in, a mathematical definition? That will show us what it means to stretch and pull these shapes?"

"It will. Luckily, we've already learned the core concept we need for that definition."

"Not me! I haven't studied topology at all!"

"But you have studied continuity."

"Really? Continuity shows up in topology?" I asked.

"Continuity is *everywhere* in topology."

"Huh."

"So to mathematically define homeomorphisms, we need to have a good mathematical grasp on just what continuity is. Do you remember the definition, Tetra? If a function $f(x)$ is continuous at $x = a$, what can we say about it?"

"Oh, okay. Give me a minute. I can do this using limits, right? I'm sure I can remember this!"

3.3.2 Continuous Functions

Five minutes later...

"Something like this?" Tetra said, showing us her definition of continuity.

A limit-based definition of continuity

A function $f(x)$ for which

$$\lim_{x \to a} f(x) = f(a)$$

is continuous at $x = a$.

"Okay," Miruka said, nodding. "Though it would be better to explicitly assert the existence of that limit."

A limit-based definition of continuity (reworded)

A function $f(x)$ for which

- a limiting value for $f(x)$ exists as $x \to a$, and
- the value of that limit equals $f(a)$

is continuous at $x = a$.

I recalled the practice problem I had worked on just the other day. "In class they teach us that $f(x)$ being continuous at $x = a$ means 'as x becomes arbitrarily close to a, $f(x)$ becomes arbitrarily close to $f(a)$.' They define limits using this 'arbitrarily close' wording, and that carries over to their definition of continuity."

"I understand the definition using limits," Tetra said in a near whisper, "but I still picture continuity not as equations, but in terms of the graph for $y = f(x)$. I mean, the graphs look different, depending on whether $f(x)$ is continuous at $x = a$, right?"

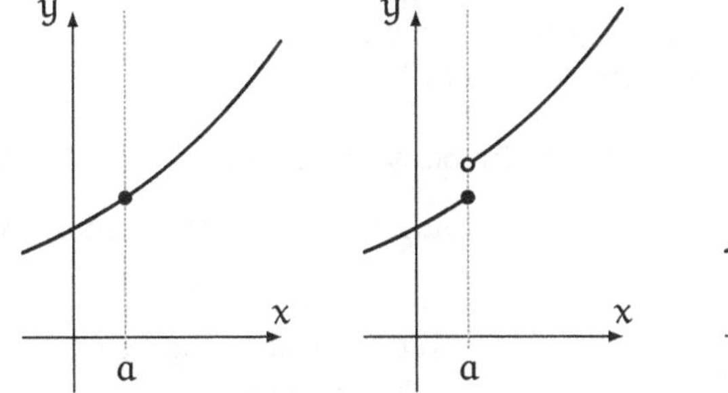

Continuous at $x = a$. Discontinuous at $x = a$

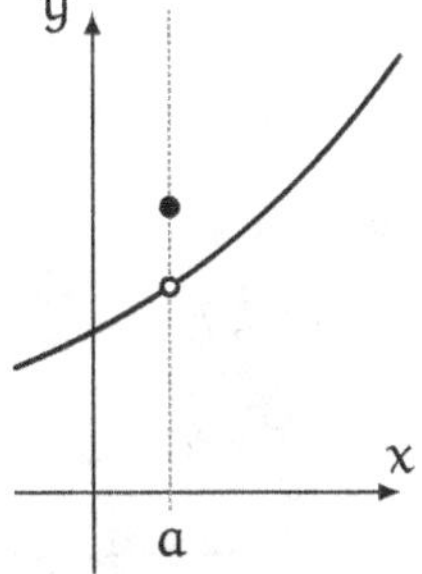

Discontinuous at $x = a$

"That's fine," Miruka said, "but let's dig a little deeper into the concept of continuity so we can arrive at our goal."

"Just what *is* our goal?" Tetra asked.

"Defining homeomorphism so we can represent 'same shapes' in topology, of course. To define homeomorphic mappings we need to define continuous mappings. We're familiar with functions that are continuous over the real numbers, so let's start digging into the concept of continuity from there. By doing so, we will also be digging into the concept of limits."

"Ah, I see! You're going to use the epsilon–delta definition of limits, right?"

"Of course. To define continuity, we need limits. But just saying things like 'arbitrarily close' is too fuzzy, so let's use logic statements instead. By representing continuity as logic statements, we can better capture its fundamental nature. Then we can use epsilon–delta to show how a function $f(x)$ can be continuous at $x = a$."

Definition of continuity (represented as logic statements)

A function $f(x)$ for which

$$\forall\, \varepsilon > 0\ \exists\, \delta > 0\ \forall\, x\ \Big[\, |x - a| < \delta \Rightarrow |f(x) - f(a)| < \varepsilon \,\Big]$$

is continuous at $x = a$.

"Remember when we went through this, Tetra?" I asked.[1] "We used this to practice reading logic statements."

"I remember! And I can still read it!"

$\forall\, \varepsilon > 0$	For every positive number ε,
$\exists\, \delta > 0$	we can choose some positive number δ
$\forall\, x\ \big[$	so that for every x
$\lvert x - a \rvert < \delta$	the distance between x and a being less than δ
$\Rightarrow$	implies that
$\lvert f(x) - f(a) \rvert < \varepsilon$	the distance between $f(x)$ and $f(a)$ is less than ε.
$\big]$	

For every positive number ε, we can choose some positive number δ so that for any x, $|x-a| < \delta \Rightarrow |f(x) - f(a)| < \varepsilon$.

[1]See *Math Girls 3: Gödel's Incompleteness Theorems*

"It took me a long time to remember all this! Even so, the epsilons and deltas still get all mixed up in my head, so I always end up going back to graphs and how they're connected when I think about continuity."

"You could just write the ε's and δ's in your graph, you know," I said. "Saying that for every ε we can choose a δ such that $|f(x) - f(a)| < \varepsilon$ means we can choose a δ such that if x is within δ of a, $f(x)$ will always be within ε of $f(a)$. That's easier to see in a graph."

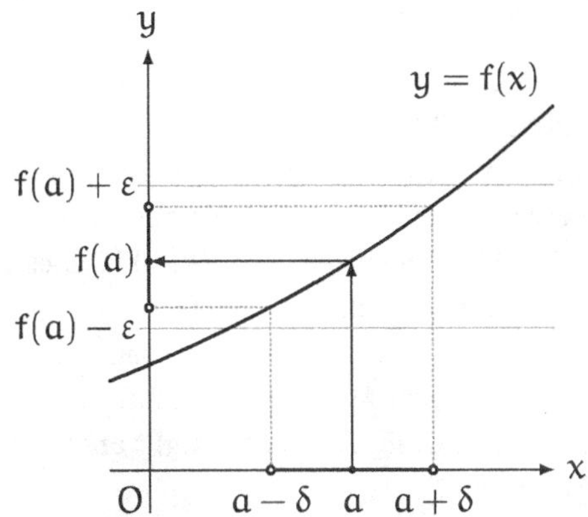

A continuous function $f(x)$ maps all points within the δ neighborhood of a to the ε neighborhood of $f(a)$.

"I guess, but still... trying to mirror the horizontal movements with the vertical movements makes me dizzy."

"Then you should write both vertically," Miruka said. "That makes it even easier to see how all points in the δ neighborhood of a get mapped to the ε neighborhood of $f(a)$."

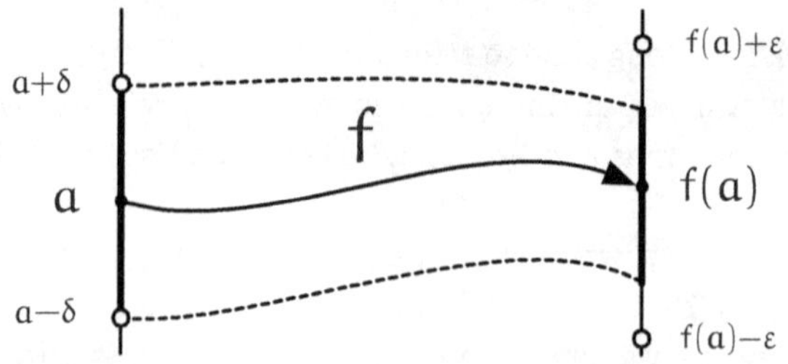

A continuous function $f(x)$ maps all points within the δ neighborhood of a to the ε neighborhood of $f(a)$.

"This area from $a-\delta$ to $a+\delta$ is what you're calling 'the δ neighborhood of a,' right?" Tetra asked.

"Yeah, except that the border isn't included," I said. "And similarly, the area between $f(a)-\varepsilon$ and $f(a)+\varepsilon$ is the ε neighborhood of $f(a)$. You're right, Miruka—writing it like this does make it easier to see how the function f maps a to $f(a)$."

"One more question. Is a 'mapping' the same as a 'function'?"

"You can generally treat them as the same thing," Miruka said. "In some situations, however, 'function' might only apply when mapping to sets of numbers."

"I like this graph," I said. "It gives a more intuitive sense of continuity, feels more like things being connected."

"How so?" Tetra asked.

"Well, because it shows that no matter how we choose ε, in other words no matter how we choose the ε neighborhood of $f(a)$, if we're careful about how we choose δ, f maps the δ neighborhood of a right into the ε neighborhood of $f(a)$."

"Hmmm..."

I turned toward Tetra. "I mean, see how even when ε is super small, if the function f is discontinuous at a, we can't always choose a good δ? Function f being continuous assures us that no matter how small ε is, we can just make δ smaller for our mapping."

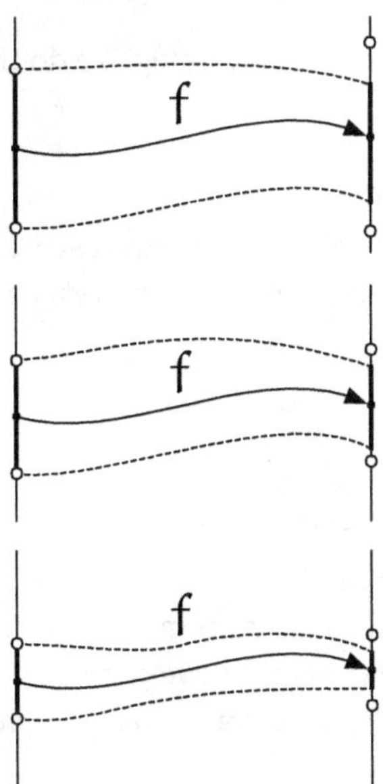

"Okay, I think I'm getting there. In other words, when we ask 'Can everybody squeeze in close to $f(a)$?', everybody near a can move in closer to $f(a)$. And if we ask 'Can everybody squeeze in *really* close to $f(a)$?', everybody close to a can move in even closer to $f(a)$."

"I guess that's the gist of it. The role of ε and δ is to be explicit in what we mean by 'close.' To close in on the fundamental nature of continuity, we need to be clear about distances. That's why the ε–δ definition of limits is useful."

Tetra nodded. "Sure. We wouldn't be able to write equations without distances."

"That's not quite true," Miruka said, slowly shaking her head. "In the ε–δ definition of limits we set some distances we call ε and δ, and use those to define continuity. Our next step is to make things far more abstract. We need to cast distances aside, and attempt a definition of continuity for a world in which distance does not exist."

"Cast distances aside? How could we possibly—"

"You just tried to explain continuity using the word 'close,' right? We need to formalize that. We need to define the 'neighbors' living in the neighborhoods you're imagining. To do that, we need to travel to space."

"To outer space?"

"To topological space."

Now things are getting really *interesting.*

Feeling a Miruka lecture coming on, Tetra and I sat back to revel in it.

3.4 Near Point a

3.4.1 Preparing for Our Journey

Miruka started her lecture.

"Okay, if we're going to leap from the world of distances to topological space, it will be useful to compare the terms we use in each. For convenience, we'll treat functions and mappings as separate things."

World of distances		Topological space
All real numbers	←----→	Sets (topological spaces)
Real number	←----→	Element (point)
Function	←----→	Mapping
Continuous function	←----→	Continuous mapping
Open interval	←----→	Connected open set
ε, δ neighborhoods	←----→	Open neighborhood

Corresponding terms.

"We've already defined continuous functions using ε–δ. Now we want to do something similar to define continuous mappings. Specifically, in the world of distances, we used an ε neighborhood belonging to $f(a)$ and a δ neighborhood belonging to a to build up the ε–δ argument. Let's follow that approach."

3.4.2 *Real-value* a *and* δ *Neighborhoods in the World of Distances*

"Let's take a closer look at a δ neighborhood for a real value a," Miruka said.

$a-\delta \qquad a \qquad a+\delta$

A δ neighborhood for a real value a.

"We can think of this as a set of real numbers."

$$\{x \in \mathbb{R} \mid a-\delta < x < a+\delta\}$$

"An interval like this, one that doesn't include its boundaries, is called an open interval. Sometimes we'll write this interval using parentheses."

$$(a-\delta, a+\delta) = \{x \in \mathbb{R} \mid a-\delta < x < a+\delta\}$$

"It would be too long to write a δ neighborhood of a as $(a-\delta, a+\delta)$, so we'll shorten that to $B_\delta(a)$."

$$B_\delta(a) = \{x \in \mathbb{R} \mid a-\delta < x < a+\delta\}$$

"We can use the same notation to show an ε neighborhood of $f(a)$ as $B_\varepsilon(f(a))$."

$$B_\varepsilon(f(a)) = \{x \in \mathbb{R} \mid f(a)-\varepsilon < x < f(a)+\varepsilon\}$$

Definition of a δ neighborhood of point a in the world of distances

$$B_\delta(a) = \{x \in \mathbb{R} \mid a-\delta < x < a+\delta\}$$

$a-\delta \qquad a \qquad a+\delta$

$B_\delta(a)$

"Just to make sure," Tetra said, "all we've done so far is decide how we're going to indicate a δ neighborhood and an ε neighborhood, right?"

Miruka nodded. "That's right. We haven't covered anything new at all, and we're still in the world of distances. Next, we want to define the concept of open sets in that world."

3.4.3 Open Sets in the World of Distances

Definition of open sets in the world of distances

Consider a subset O of the set $\mathbb{R}$ of all real numbers. O is called an *open set* if for every real number a belonging to O, there exists some ε neighborhood of a in O.

$$O \text{ is an open set} \iff \forall a \in O \ \exists \varepsilon > 0 \left[B_\varepsilon(a) \subset O \right]$$

"This is our definition of open sets in the world of distances," Miruka said. "I think it's pretty clear, but let's test our understanding.

"Say we have two real numbers u and v, with $u < v$. Is the open interval $(u, v) = \{x \in \mathbb{R} \mid u < x < v\}$ an open set? What do you think, Tetra?"

"I'm... um... not sure," Tetra said.

Miruka turned to me. "What do you think?"

"I think it is," I replied. "For any real number a in $u < a < v$, we can take an $\varepsilon > 0$ small enough that we don't leave the open interval (u, v). So I think that satisfies the definition of open sets here. As a graph, it would look like this."

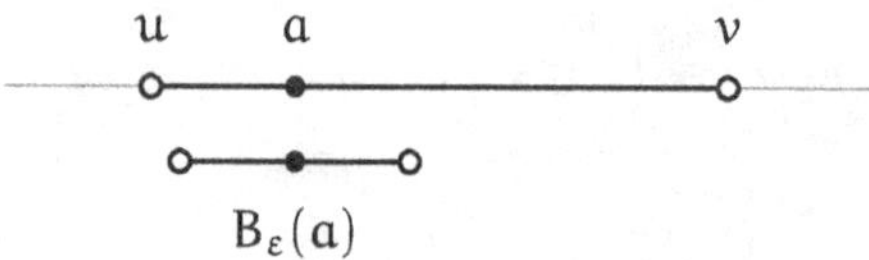

Open interval (u, v) and $B_\varepsilon(a)$ (an ε neighborhood of a).

"Very good. If there exists an ε such that $B_\varepsilon(a) \subset (u, v)$, we can say that the open interval (u, v) is an open set."

"Okay, I see what this definition is saying now," Tetra said. "So specifically, we just need an ε that's smaller than both $a - u$ and $v - a$, right?"

"That's right," I said.

"Okay, I'm good now. I'm still not sure what this definition is for, though. If we can make really small ε's for any set, doesn't everything become an open set?"

"No," Miruka snapped. "You've forgotten *closed* sets $[u, v]$. The set of all real numbers x where $u \leqslant x \leqslant v$ is *not* an open set."

$$[u, v] = \{x \in \mathbb{R} \mid u \leqslant x \leqslant v\}$$

"Why not, Tetra?"

"Because, um... because u and v would be exceptions?"

"Good. No matter how small you make $\varepsilon > 0$, an ε neighborhood for point u would extend outside of $[u, v]$. So the closed interval $[u, v]$ is not an open set."

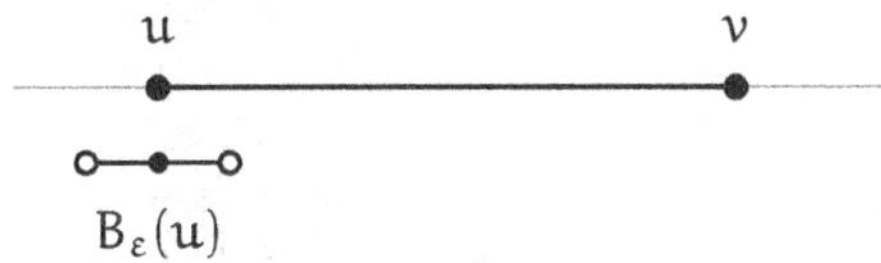

Closed interval $[u, v]$ and $B_\varepsilon(u)$ (an ε neighborhood of u).

"Okay, hang on. I see how a closed interval isn't an open set. But then, isn't an open set the same as an open interval? I still don't see why we're defining open sets."

"Open sets and open intervals are different things," Miruka said. "We can create a union of sets from multiple open intervals. For example, this union of open intervals (u, v) and (s, t) is an open set."

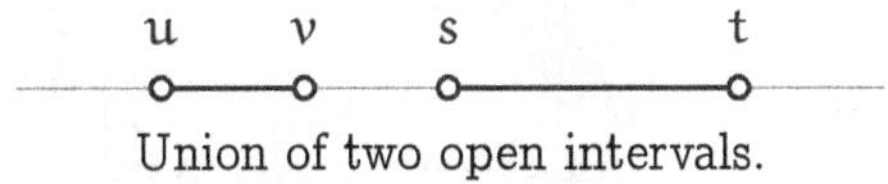

Union of two open intervals.

"Ah, right. Gotcha," Tetra said.

3.4.4 Properties of Open Sets in the World of Distances

"So now we've used ε neighborhoods to define open sets," Miruka continued. "Note that we're still in the world of distances. We want to prepare ourselves for leaving this world, though, so our next step is to investigate the properties of these open sets we've just defined. In particular, we want to focus on their properties *as sets*. There are four we should discuss."

Properties of open sets in the world of distances

Property 1: The set $\mathbb{R}$ of all real numbers is an open set.

Property 2: The empty set is an open set.

Property 3: The intersection of two open sets is an open set.

Property 4: The union of any number of open sets is an open set.

"Interesting," I said.

"Property 1 here says the real numbers make an open set. But of course they are, since we can say that $B_\varepsilon(a) \subset \mathbb{R}$ for any real number a we choose, taking any positive real number as ε."

"Does Property 2 really hold?" Tetra asked. "I mean, the empty set doesn't have any elements! How can we take an ε neighborhood for a point a?"

"It holds," Miruka said. "Think about whether this holds for a set $O = \{\}$."

$$\forall a \in O\ \exists \varepsilon > 0 \left[B_\varepsilon(a) \subset O \right]$$

"It has to, right? If that's hard to see, think about this, using the same values."

$$\neg \left[\exists a \in O\ \underset{\sim\sim\sim\sim\sim\sim\sim\sim\sim\sim\sim\sim}{\forall \varepsilon > 0 \left[B_\varepsilon(a) \not\subset O \right]} \right]$$

"Since O is the empty set, there is no a that can satisfy the underlined part here, no matter what conditions you put on it. Since O is the empty set, there's no way to set up a counterexample."

"I see."

"Property 3 says the intersection of two open sets is an open set, and again, this is easy to see," Miruka continued. "This isn't limited to just two sets, but any finite number of them. However, an intersection of infinitely many open sets isn't necessarily an open set. For example, consider the sequence of open sets $B_{\frac{1}{n}}(0)$. Then all its intersections would be this."

$$\bigcap_{n=1}^{\infty} B_{\frac{1}{n}}(0) = \{0\}$$

"But the set $\{0\}$ having only 0 as an element isn't an open set."

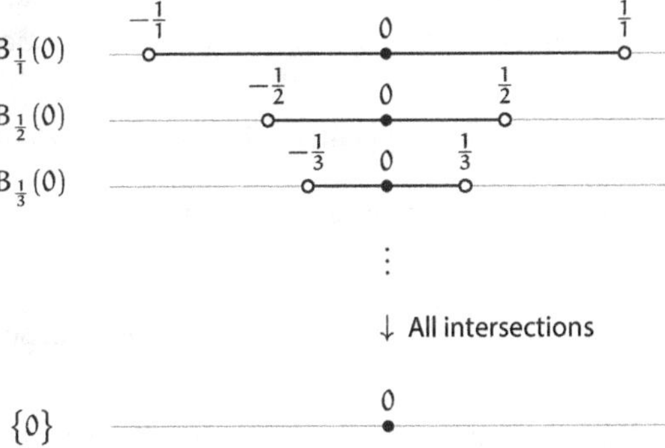

"Only finitely many open sets," I said. "Duly noted."

Miruka went on. "Property 4 says a union of any number of open sets is an open set. In this case we can handle infinitely many open sets. No matter what point a belonging to a union of sets we consider, a must have belonged to one of the open sets before their union, so an ε neighborhood of point a must exist.

"And that's the last one. We've shown that these four properties hold for open sets in the world of distances."

Tetra raised her hand. "Sorry. I understand each of these properties one-by-one, but I'm still a little bit lost. Where exactly are we going with all this?"

"Yeah, I'm kinda lost myself," I admitted. "You said our goal is to move from 'the world of distances' to 'topological space,' right?"

"Hmph. Okay then, let's scout out our route, shall we?" Miruka said.

3.4.5 The Route from the World of Distances to Topological Space

"We want to define homeomorphisms," Miruka began. "To do that, we need continuous mappings. We know about continuous functions in the world of distances, so let's use those as a model for defining continuous mappings between topological spaces.

"One problem, though—to transition between these two worlds we need to get rid of distances, but without distances, we can't create δ neighborhoods or ε neighborhoods. Without those, we can't use epsilon–delta to show continuity. A problem indeed.

"So what we do is create the concept of open sets in the world of distances, because that will allow us to create the concept of open neighborhoods, which we can use in place of δ neighborhoods and ε neighborhoods.

"Here's what we've done so far in the world of distances."

> In the world of distances: For the set $\mathbb{R}$ of all real numbers . . .
>
> - we used ε neighborhoods to define open sets, and
> - we confirmed the properties of open sets.

"We're nearly ready to blast off from the world of distances into topological space. Along the way we'll be discarding the concept of distances, so in topological space we can't use ε neighborhoods. Which means we can't use our definition of open sets, because that relied on ε neighborhoods. So what *can* we use?

"Well, for a start we can reinterpret the *properties* of open sets we saw in the world of distances as *axioms* of open sets. In topological space, we can define open sets as if their axioms came raining down from the heavens. Like this."

> In topological space: For any set S . . .
>
> - we use the open set axioms to define open sets, and

· we use open sets to define something called open neighborhoods.

"Once we've obtained open neighborhoods in topological space, we're just one step from defining continuous mappings. Are you starting to see the path we're going to follow?"

"I am!" Tetra said, raising her hand. "But is it okay if I write this along with the terms you taught us? Does a map of our journey look something like this?"

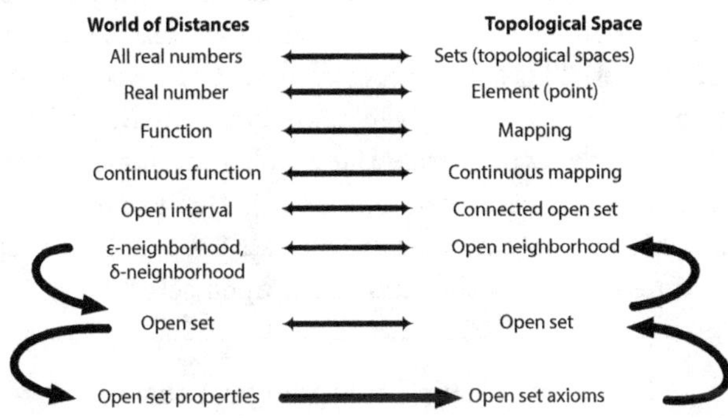

A map of our journey

"Well done," Miruka said.

I was surprised to see how well Tetra had followed Miruka's lecture so far, well enough to just write it out as a "map." She had some special power to... what? Find consistency in what she was told? See the big picture in things?

"So I guess our next job is to take a look at the axioms for open sets?" Tetra said.

"That's right," Miruka said. "That's what takes us into topological space."

3.4.6 Open Set Axioms in Topological Space

"Consider a set S," Miruka began. "We want to define an open set for this S. The question is, how?

"To start, let's gather up a set of subsets of S and call that set $\mathbb{O}$. Since $\mathbb{O}$ is a set of subsets of S, any element $O \in \mathbb{O}$ will be a subset of S. In other words, $O \subset S$. We will call $\mathbb{O}$ the set of all open sets of S, so if $A \in \mathbb{O}$ then A is an open set, otherwise it isn't.

"Also, $\mathbb{O}$ is a set of subsets of S, but we can't just arbitrarily choose these subsets. They all have to satisfy the axioms for open sets, which we'll get to soon.

"When $\mathbb{O}$ satisfies the axioms for open sets, we say that $\mathbb{O}$ endows the set S with a topological structure. We can also say $\mathbb{O}$ has given S a topology.

"The combination of S and $\mathbb{O}$ together, which we'll write as $(S, \mathbb{O})$, is called a topological space. And this isn't limited to topological structures, but the set S on which the structure is built is called the underlying set. There isn't necessarily just one way to determine the set of open sets $\mathbb{O}$ for a given underlying set S, and changing how we determine $\mathbb{O}$ will change the topological space. That's why we consider S and $\mathbb{O}$ as a pair, like $(S, \mathbb{O})$. But sometimes, if $\mathbb{O}$ is understandable by context, we can just call S a topological space.

"Okay, time to look at the axioms for open sets."

Open set axioms (in topological space)

Axiom 1. Set S is an open set ($S \in \mathbb{O}$).

Axiom 2. The empty set is an open set ($\{\} \in \mathbb{O}$).

Axiom 3. The intersection of two open sets is an open set (if $O_1 \in \mathbb{O}$ and $O_2 \in \mathbb{O}$, then $O_1 \cap O_2 \in \mathbb{O}$).

Axiom 4. The union of any number of open sets is an open set. In other words, letting Λ be a given index set, if $\{O_\lambda \in \mathbb{O} \mid \lambda \in \Lambda\}$, then $\bigcup_{\lambda \in \Lambda} O_\lambda \in \mathbb{O}$.

"I think you can see how these four open set axioms correspond to the four characteristics of open sets we saw in the world of distance. But don't forget—we've already discarded the concept of 'distance,' so we have to pretend like we don't know anything about the properties of open sets there. So we'll treat these open set axioms as if they just fell from the sky. These are axioms, so there's no need to prove them. Think of them as... promises. Things we'll need to support our argument. We want to think about what these axioms allow us to say. But before that, let's be sure we understand what these axioms require."

Miruka started talking more quickly. She was obviously having fun.

"Axiom 1 tells us set S is an open set. In other words, when we determine $\mathbb{O}$, we must have S as an element of $\mathbb{O}$. Otherwise we can't call $(S, \mathbb{O})$ a topological space. That's what these axioms from the heavens tell us.

"Same thing for the other axioms. Axiom 2 demands that the empty set be an open set. Axiom 3 says the intersection of two open sets is an open set. By repetition, we can derive that the intersection of any finite number of open sets is an open set. Finally, Axiom 4 tells us the union of any number of open sets is an open set. No need for limiting ourselves to finite numbers here."

"What's this 'index set' in Axiom 4?" Tetra asked.

"An index set is just a set of indices λ for O_λ, but maybe that doesn't clear things up much. Axiom 4 is the only one that looks confusing, but that's just because it's being careful to handle infinities. Let's take a closer look at it, step-by-step. First, if you have finitely many open sets $O_1, O_2, \ldots, O_n$, the open set axioms say their union must in turn be an open set. In that case, the index set is $\Lambda = \{1, 2, \ldots, n\}$."

$$\bigcup_{\lambda \in \Lambda} O_\lambda = O_1 \cup O_2 \cup \cdots \cup O_n \in \mathbb{O}$$

"But with Axiom 4 we can consider unions of infinitely many open sets, because it says the union of open sets $O_1, O_2, \ldots$ is also an open set. In this case, the index set is $\Lambda = \{1, 2, \ldots\}$."

$$\bigcup_{\lambda \in \Lambda} O_\lambda = O_1 \cup O_2 \cup \cdots \cup O_n \cup \cdots \in \mathbb{O}$$

"So here's the problem. It feels like $1, 2, \ldots$ should be enough subscripts, but that will only work when we're dealing with a countable set, namely the set of all positive integers. Like we saw when we learned about Cantor's diagonal argument[2], there are different kinds of infinite sets. For example, we could also use elements of an uncountable set like the set of all real numbers as our indices. That's why Axiom 4 goes to the trouble of setting up this index set Λ and writes unions using its elements. We can also write Axiom 4 without using an index set, like this."

$$\mathbb{O}' \subset \mathbb{O} \Rightarrow \bigcup_{O \in \mathbb{O}'} O \in \mathbb{O}$$

"Ah, interesting," I said.

"So. Where to next?" Miruka asked.

"A definition for open neighborhoods!" Tetra said, referring to the map she'd prepared.

"Well let's get to it, then."

3.4.7 Open Neighborhoods in Topological Space

"So here we are, in topological space," Miruka said. "We've defined open sets. Given a point a belonging to S, an open set having a as an element is called an open neighborhood of a. That's what it means to be 'near' point a."

Definition of an open neighborhood of point a in topological space

An open set having a point a as an element is an open neighborhood of a.

Tetra raised her hand. "One question. Is there any way you can, like, visualize this?"

"Sure. You can imagine the open neighborhood of a as looking something like this."

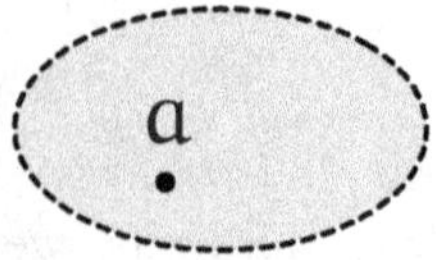

The open neighborhood for a point a.

[2]See *Math Girls 3: Gödel's Incompleteness Theorems.*

"Simple enough, I guess."

"Be careful, though. This looks like a two-dimensional shape, but it isn't. It's just a concept diagram. In figures you'll often see the open neighborhood of a point a written as a dot surrounded by a dashed line, but this is just to remind you that back in the world of distances, an open set doesn't include the border around it."

Tetra looked through her notes. "In the world of distances, we used $B_\delta(a)$ to write a δ-neighborhood of a real number a. Can we write something like $B(a)$ to mean an open neighborhood of a in topological space?"

"No, we can't. Because there isn't just one open neighborhood for point a. Of course, it would be inconvenient to not be able to handle things using equations, so let's use $\mathbb{B}(a)$ to describe the set of all open neighborhoods for a. In other words, we're using $\mathbb{B}(a)$ to denote the set of all open sets having a as an element."

Set of all open neighborhoods of a (in topological space)

We write $\mathbb{B}(a)$ to denote the set of all open neighborhoods of a, where

$$\mathbb{B}(a) = \{O \in \mathbb{O} \mid a \in O\}.$$

"So $\mathbb{B}(a)$ means 'near a,' right?" Tetra asked.

"Wrong. $\mathbb{B}(a)$ is *all* of the 'near a's.' Any one 'near a' is an element in $\mathbb{B}(a)$. Like this."

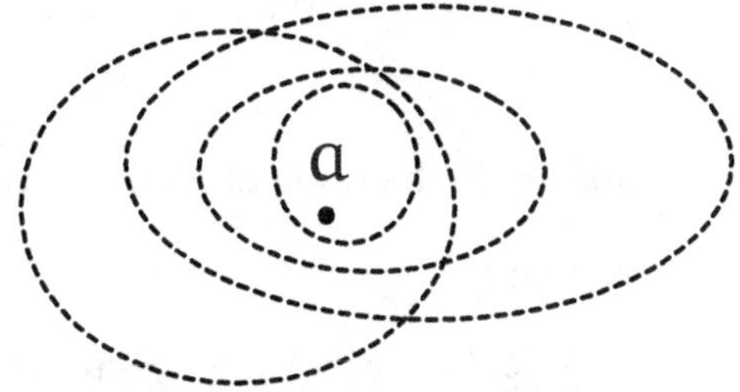

$\mathbb{B}(a)$ is the set of all open neighborhoods for a point a.

"Ah, okay! I get it now."

"Now we can define continuous mappings," I said.

"We can indeed," Miruka replied, "now that we've defined open neighborhoods in topological space."

3.4.8 Continuous Mappings in Topological Space

"So now we can finally define continuous mappings. But first, let's review continuous functions in the world of distances."

Definition of continuity in the world of distances

A function $f(x)$ satisfying

$$\forall \varepsilon > 0 \ \exists \delta > 0 \ \forall x \ \Big[|x - a| < \delta \Rightarrow |f(x) - f(a)| < \varepsilon \Big]$$

is continuous at $x = a$.

"This definition is written using distances, but we want to rewrite it using δ- and ε-neighborhoods. For example, we can write this."

$$|x - a| < \delta$$

"This means the distance between x and a is less than δ, so it's equivalent to saying x is within a δ-neighborhood of a."

$$|x - a| < \delta \iff x \in B_\delta(a)$$

"In the same way, we can say this."

$$|f(x) - f(a)| < \varepsilon \iff f(x) \in B_\varepsilon(f(a))$$

Definition of continuity in the world of distances (reworded)

A function $f(x)$ satisfying

$$\forall \varepsilon > 0 \ \exists \delta > 0 \ \forall x \ \Big[x \in B_\delta(a) \Rightarrow f(x) \in B_\varepsilon(f(a)) \Big]$$

is continuous at $x = a$.

"So we've seen the definition of continuity in the world of distances. Let's take a look at that definition in topological space."

Definition of continuity in topological space

A mapping $f(x)$ satisfying

$$\forall E \in \mathbb{B}(f(a)) \; \exists D \in \mathbb{B}(a) \; \forall x \; \Big[x \in D \Rightarrow f(x) \in E\Big]$$

is continuous at $x = a$.

Miruka nodded. "A perfect translation."

"Very interesting!" I said.

"Hold up," Tetra said. "I need to read this carefully."

$\forall E \in \mathbb{B}(f(a))$	For every open neighborhood E of $f(a)$
$\exists D \in \mathbb{B}(a)$	we can select an appropriate open neighborhood D of a
$\forall x \; \Big[$	then for any x, we have chosen D so that
$x \in D$	if x belongs to D
$\Rightarrow$	then
$f(x) \in E$	$f(x)$ belongs to E.
$\Big]$	Hooray!

"So we're saying no matter which E we choose, we can also choose some appropriate D." I said. "An appropriate D is one where so long as x belongs to D, then $f(x)$ gives some point belonging to E. And we can choose some D like that. In the world of distances we had 'there exists some δ satisfying...', but in topological space we have 'there exists some open neighborhood D of a satisfying...'. So yeah, the distances are gone and...yep, looks like you've rewritten epsilon–delta!"

"All distances fade away in topological space," Miruka said. "Note how we've defined continuity using $\in$ from sets, $\mathbb{B}(a)$ and $\mathbb{B}(f(a))$ from topology, and $\forall$, $\exists$, and $\Rightarrow$ from logic."

"Give me a sec to compare the two," Tetra said, pulling two sheets of paper next to each other.

Definitions of continuity in the world of distance and topological space

$$\forall \varepsilon > 0 \quad \exists \delta > 0 \quad \forall x \left[x \in B_\delta(a) \Rightarrow f(x) \in B_\varepsilon(f(a)) \right]$$

$$\forall E \in \mathbb{B}(f(a)) \; \exists D \in \mathbb{B}(a) \; \forall x \left[\quad x \in D \Rightarrow f(x) \in E \quad \right]$$

"They are similar, I guess," she said. "But still, I'm having trouble keeping them separate."

Miruka sketched a figure and pushed it toward Tetra.

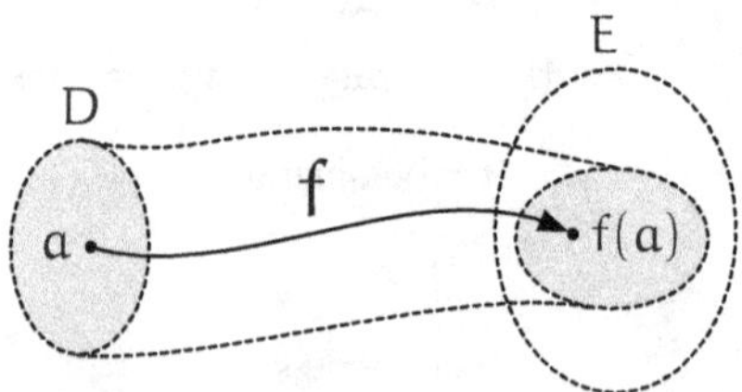

Mapping f maps all points in an open neighborhood D of a to an open neighborhood E of f(a).

"Oh, ho!" Tetra said, flipping back through her notebook. "Sure enough, that's a lot like this other image for continuity in the world of distances!"[3]

I nodded. "If the open neighborhood E of f(a) gets smaller, we can just choose a smaller open neighborhood D of a, just like choosing a smaller δ to make up for a smaller ε."

[3] See p. 88.

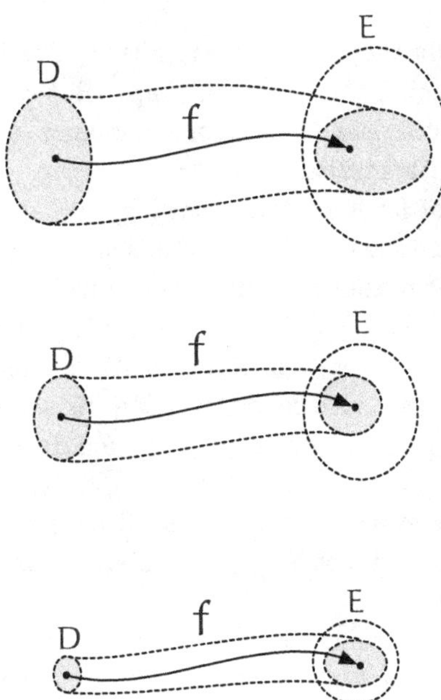

"We can set the open neighborhood E however we like," Miruka said. "It doesn't matter how small E becomes, because as long as it still contains $f(a)$ we can still whip out an open neighborhood D of a for that particular E, one where the mapping f moves every point belonging to D to somewhere in E. And we're calling an f like that continuous at $x = a$. Also, we'll call a mapping that's continuous at every point a 'continuous mapping.'"

"So that's the abstracted definition of continuity," I said.

"I see," Tetra said, though somewhat hesitantly and furrowing her eyebrows. "But Miruka, I understand how we were able to get rid of distances, and how we were able to define continuous mappings in topological space in a way that's similar to good old epsilon–delta, but... why? I can't help but feel like we've just made continuous mappings even harder to understand than they were before. Isn't it easier to just think in terms of graphs, and say there's continuity if the graph isn't broken, or discontinuity if it is?

"Sounds like you're still too fixated on the real number line," Miruka said. "What we're doing here should expand that concept. Namely, that continuous mappings aren't limited to real numbers, though admittedly they're where we initially developed our image of continuous functions, like in your graph. But still, I want you to see that we can still define continuous mappings, even without real numbers, even without distances."

Tetra and I sat for a moment, chewing on that.

Tetra broke the silence. "There's still something difficult in that. Can you give me an example that doesn't involve real numbers?"

"Hmmm, let's see... Well, the face cards in a deck of playing cards are the jack, queen, and king, right? So consider a set $S = \{J, Q, K\}$ having those three elements. Then we can define a topological structure $\mathbb{O}$ on S as its underlying set, and define our continuous mapping."

Tetra's face screwed up in confusion. "I have no idea..."

"Tell you what, let's save that for later. I want to move on to homeomorphisms."

"Hang on," Tetra said, her hand shooting up. "One more thing that's been bothering me. It seems like this whole thing with open neighborhoods is just a matter of hiding distances in $\mathbb{B}(a)$. To decide who's a neighbor without using any distances, all we did was define an open neighborhood for a as a set of points near a. But if things are so easy, it seems like we can just get away with anything, just by defining whatever we want! Is that really math?"

Miruka smiled. "Now you're starting to sound like me," she said, causing Tetra to shoot me a glance. "But don't forget, there are axioms for open sets, placing restrictions on what they can be. So it's not as if we just pulled our definition of open neighborhoods out of thin air."

"Oh, the open set axioms, right."

"A topological space is an underlying set S with a topological structure," Miruka said, rubbing her left arm with her right hand. "It's just like when we create the structure of a group. Just as we define groups using the group axioms, we define topological spaces using the open set axioms. Our definition of continuous mappings on a topological space is exactly like the epsilon–delta definition of continuous functions that we know so well, except for the fact that there are no real numbers involved. No epsilons or deltas or limits or absolute values, either. But even so, you should be able to see how the concept is a good description of what it means to be continuous."

Tetra's hand shot up once again. I wondered how many times it had gone up that day.

"Okay, just one more thing I want to be sure of. So we're free to create a specific open set however we want, right?"

"Sure, so long as the open set axioms are satisfied," Miruka said. "Interestingly, we can bring back those distances that we started out by getting rid of. I mean, if we make sure the open sets we defined using ε-neighborhoods in the world of distances satisfy the open set axioms in topological space—which of course they will—then we can consider the set of all real numbers from the world of distances as a topological space. Lisa would probably say we're using the distances defined by absolute values of real numbers to define open sets."

"A definition using absolute values..." Tetra muttered.

"Oh, that reminds me!" I said. "Absolute values!"

Miruka frowned. "We've been talking about them for some time, you know."

"No, no, something else. Absolute values, right. That's what we talked about that first day I met you, Tetra[4]. In the auditorium, remember?"

Tetra gave a slight nod, but otherwise ignored my outburst. *Sigh.*

"Actually, thinking of groups helps me see why we want to use axioms to define topological spaces," she said. "When you guys first taught me about groups[5], I was totally lost. I thought math was all about the numbers and calculations I'd learned in school, so when I got hit with something as abstract as groups, it was kind of a shock. But you showed me how so long as the group axioms are satisfied, we can use them to perform abstracted calculations. Oh, and the same thing with ladder networks[6]. Those don't seem to have anything to do with calculations, but we saw how you can connect them together and consider that as multiplication, making them form groups."

"We also did something similar with vector spaces," I said. "Matrices, rational numbers, and algebraic expressions are all different

[4]See *Math Girls*, section 2.5.2.

[5]See *Math Girls 2: Fermat's Last Theorem.*

[6]See *Math Girls 5: Galois Theory.*

things, but by using the vector space axioms, we saw how they can be organized from the same viewpoint."

"And therein lies the power of mathematics and logic," Miruka said. "We just have to prepare a set of axioms that give some minimum requirements. Increasing the level of abstraction may make things harder to understand, but that's what gives us our wings. It's how we notice that things so seemingly different at face value actually have the same structure, or that we can provide them with the same structure."

"Huh," I said. "So in the past we've equipped sets with the structure of groups, and with the structure of vector spaces. Now we're doing the same thing, using the open set axioms to equip sets with this structure called a topology."

"Namely, we defined open neighborhoods using this abstract idea of being 'real close,' and that allowed us to incorporate the concept of continuous mappings into topological space."

"This is really cool. I knew topology was something about soft, squishy forms. But I didn't realize we could use all these familiar concepts to provide that soft squishiness."

Tetra placed a hand on either side of her head. "My mind is still spinning a bit, but I think I've got the general idea. So we've obtained continuous mappings. Now we can finally move on to talking about same shapes, in other words homeomorphisms?"

"Wow, I'd almost forgotten about that!"

"We've defined topological spaces," Miruka said, "and we've defined continuous mappings, so yes, now we can define homeomorphisms. And we'll see that we can describe them much more simply than continuous mappings."

3.4.9 Homeomorphisms

Definition of homeomorphism

Let X and Y be topological spaces, and let f be a bijective function mapping X to Y. If both f and its inverse f^{-1} are continuous, f is called a *homeomorphism*. Further, if there exists a homeomorphism from X to Y, we say that X and Y are *homeomorphic*.

"A topological space is a set that has been given a topological structure," Miruka began. "In other words, you can think of it as a set for which open sets have been defined. By defining open sets, we can define open neighborhoods. By defining open neighborhoods, we can define continuous mappings. If a continuous mapping f is bijective, its inverse mapping f^{-1} exists. Then, we make this definition."

> Given a bijective mapping f between topological spaces X and Y, if f and f^{-1} are continuous, then X and Y are homeomorphic.

"Hold up," Tetra said. "Remind me what a bijective mapping is?"

"A mapping where for any element x in X there is exactly one element y in Y such that $y = f(x)$, and conversely for any element y in Y there is exactly one element x in X such that $y = f(x)$. For any bijection $y = f(x)$, we can define its inverse mapping $x = f^{-1}(y)$."

"And that f is a homeomorphism, right?" I said. "We have a bijection that's continuous both ways, so we can create a correspondence between the two topological spaces."

"That's right. We can treat two homeomorphic topological spaces as being the 'same,' topologically speaking."

Tetra remained silent.

I let out a low groan as I tried to do some mental organization. "So we have all these sets that seemingly have nothing to do with each other, and we apply a topology to it—in other words, we find open sets and open neighborhoods—and that lets us define continuity. Then we use that continuity to define homeomorphism, and we can make that definition without ever considering 'actual' distances... Hang on, since we've defined continuity, we can use limits, right? Doesn't that mean we can define not just continuity, but also derivatives?"

"Not quite. To define derivatives we need not just topological structures, but differentiable structures. Not that that's impossible—when 'sameness' extends to differentiation, we call that diffeomorphism. Actually, Poincaré's use of the word homeomorphism included diffeomorphism. But anyway, now that we've defined homeomorphism, we can notice something that's of high interest in topology..."

3.4.10 Invariance

"We've defined homeomorphism," Miruka said. "That means we've defined what it means for things to have the 'same shape' in topology. Homeomorphisms are very important. Topology is highly interested in quantities that do not change under homeomorphism, in other words invariant quantities. In topology, these are unsurprisingly called topological invariants."

"Because things that don't change are worth naming, right?" Tetra said.

"They are indeed. When we stretch or otherwise deform a figure, it looks like its shape has changed. But some properties will remain the same, and it's those properties we want to pay attention to. If no matter how much we stretch or shrink a form there exists a homeomorphism between its deformations, then they are homeomorphic. In a topological sense, it's still the same shape. So we're studying those quantities that stay the same between the original shape and its deformation, in other words its topological invariants."

"The Bridges of Königsberg!" I half shouted. "The problem where we thought about whether a given graph can be traversed. It's the same thing! Because no matter how much we stretch or shrink the edges, that property will remain the same!"

"That property has a name, by the way—an Eulerian walk. Whether a graph has an Eulerian walk is an invariant quality of graph isomorphisms, so that's a little different from topological invariants from homeomorphisms. Of course, you're right in the sense that both are invariant quantities."

"Ah, okay. But, look who popped up again!"

"Euler is everywhere!" Tetra said.

"He was quite the mathematician," Miruka agreed, a starry look in her eyes.

3.5 Near Tetra

On our way home, Miruka split off on a different route, saying she wanted to run by the bookstore. Tetra and I continued toward the train station, I doing my best to walk slowly to match her pace. My mind was filled with invariant properties and topological invariants and what these things we'd been studying might teach us about the true nature of forms.

From that my mind turned to the true nature of family. Families seemed to be ever-changing, yet unchanging nonetheless. What invariant properties might be behind that? What properties made a family a family?

"So Miruka says she's going back to America next week," Tetra said, interrupting my musings.

"Oh yeah?"

Miruka had already decided to skip the whole college admissions thing in Japan; she would be moving on to a university in the United States. She was already starting to see the form of her future. Me? Not so much. My mind was consumed with test scores and GPAs and studying for entrance exams. Finals in the fall. In winter, mock exams to see what schools I could likely get into. After that, the real deal. I could practically hear the countdown ticking.

"Sorry to always be taking up your time," Tetra said.

"I don't mind. Helps me clear my head."

"All this talk of same shapes is interesting, isn't it? Congruence, similarity, homeomorphisms... Who'd have thought there are so many ways to be the same!"

"Right? I wonder if today you're really the same Tetra you were yesterday?" I said, remembering our conversation from the day before.

"Same as always!" she cheerily returned, turning to face me. "Oh, but not for long! I'm about to become a new Tetra. Even I can't stay the same forever."

"Huh. So what's gotten into you?"

"Um... No, it's still a secret. I've decided on a name, though. *Eulerians*!"

"A name? A name for what?"

"That's the secret," she said, holding a finger to her lips.

Every conclusion presumes premises. These premises are either self-evident and need no demonstration, or can be established only if based on other propositions; and, as we cannot go back in this way to infinity, every deductive science, and geometry in particular, must rest upon a certain number of indemonstrable axioms.

HENRI POINCARÉ
(TRANS. BY W.J. GREENSTREET)
Science and Hypothesis

Addendum: Creating Topological Spaces and Continuous Mappings with Playing Cards

Topological Spaces

The following describes how to create topological spaces and continuous mappings using a jack (J), queen (Q), and king (K) from a deck of playing cards.

Define an underlying set S as

$$S = \{J, Q, K\}.$$

Then defining for this S the set of all open sets $\mathbb{O}$ as, for example,

$$\mathbb{O} = \big\{\{\ \}, \{Q\}, \{J, Q\}, \{Q, K\}, \{J, Q, K\}\big\},$$

$\mathbb{O}$ satisfies open set axioms 1–4 (see p. 98).

Specifically, $\mathbb{O}$ satisfies axiom 1, because $S = \{J, Q, K\} \in \mathbb{O}$.

$\mathbb{O}$ satisfies axiom 2, because $\{\ \} \in \mathbb{O}$.

$\mathbb{O}$ satisfies axiom 3, because we can confirm by brute force that the intersection of two open sets is an open set, as $\{J, Q\} \cap \{Q, K\} = \{Q\} \in \mathbb{O}$ and $\{\ \} \cap \{J, Q, K\} = \{\ \} \in \mathbb{O}$ and so on.

$\mathbb{O}$ satisfies axiom 4, because we can confirm by brute force that the union of two open sets is an open set, as $\{\ \} \cup \{Q, K\} = \{Q, K\} \in \mathbb{O}$ and $\{J, Q\} \cup \{Q, K\} = \{J, Q, K\} \in \mathbb{O}$ and so on. $\mathbb{O}$ has finitely many elements, so even when taking an arbitrary number of unions, we can reduce that to a union of two sets.

Therefore, $\mathbb{O}$ gives set S a topological structure, and $(S, \mathbb{O})$ is a topological space.

> Open neighborhoods of J are those open sets having J as an element, namely $\{J, Q\}$ and $\{J, Q, K\}$. Open neighborhoods of Q are those open sets having Q as an element, namely $\{Q\}$, $\{J, Q\}$, $\{Q, K\}$, and $\{J, Q, K\}$. Open neighborhoods of K are those open sets having K as an element, namely $\{Q, K\}$ and $\{J, Q, K\}$.

Continuous Mapping f

Define a mapping f from S to S by

$$f(J) = K, f(Q) = Q, f(K) = J.$$

f is a continuous mapping on the topological space $(S, \mathbb{O})$, because for any point a in S and for any open neighborhood E of $f(a)$ there exists an open neighborhood D of a, so

$$\forall x \left[x \in D \Rightarrow f(x) \in E \right] \qquad \cdots\cdots (\heartsuit)$$

holds.

As an example, the following shows that mapping f is continuous at J.

- If $E = \{J, Q, K\}$ is a neighborhood of $f(J) = K$, we can choose $D = \{J, Q, K\}$ as a neighborhood of J. Since $f(J), f(Q)$, and $f(K)$ are in E, ♡ holds. We could also have chosen $D = \{J, Q\}$.
- If $E = \{Q, K\}$ is a neighborhood of $f(J) = K$, we can choose $D = \{J, Q\}$ as a neighborhood of J. Since $f(J)$ and $f(Q)$ are in E, ♡ holds.

For $f(J) = K$ there are two open neighborhoods $\{J, Q, K\}$ and $\{Q, K\}$, so mapping f is continuous at J.

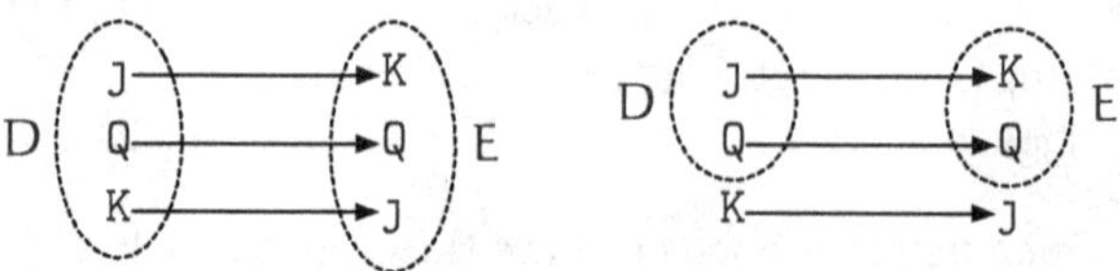

We can also confirm that mapping f is continuous at Q and K in a similar manner.

Discontinuous Mapping g

Define a mapping g from S to S by

$$g(J) = Q, g(Q) = K, g(K) = J.$$

Here, g is continuous at Q, but discontinuous at J and K. As an example, g is discontinuous at J because for an open neighborhood E for $g(J)$, we cannot always select an open neighborhood D for J such that

$$\forall x \left[x \in D \Rightarrow g(x) \in E \right] \qquad \cdots\cdots(\Diamond)$$

holds. Concretely, $E = \{Q\}$ is an open neighborhood for $g(J) = Q$, but there exists no open neighborhood D for J in $\mathbb{O}$ satisfying Eq. ($\Diamond$). There are two open neighborhoods $\{J, Q\}$ and $\{J, Q, K\}$ for J, but when letting either $D = \{J, Q\}$ or $D = \{J, Q, K\}$, for element Q in D we have $g(Q) = K$, so $g(Q)$ is not an element in E.

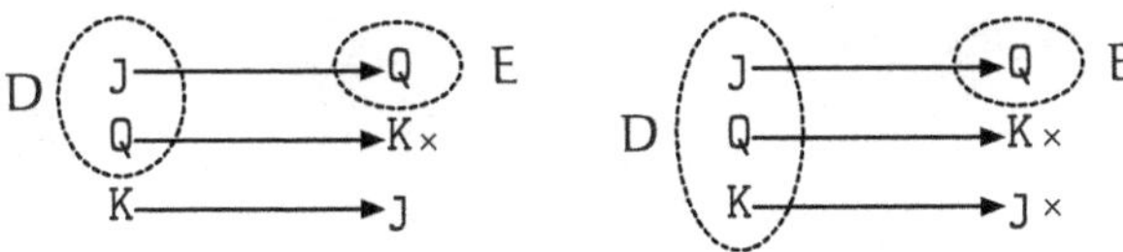

We can also confirm that mapping g is continuous at Q and discontinuous at K in a similar manner.

CHAPTER 4

Non-Euclidean Geometry

> If two lines are drawn which intersect a third in such a way that the sum of the inner angles on one side is less than two right angles, then the two lines inevitably must intersect each other on that side if extended far enough.
>
> EUCLID
> *Elements* [26]

4.1 Spherical Geometry

4.1.1 *Shortest Path on a Globe*

"Miruka's in the States *again*?" Yuri asked.

It was a Saturday, with Yuri in my room as usual. I was studying at my desk, and she was stretched out on my floor, reading a book.

"Looks like it," I said. "I think she'll be back next week, though."

Miruka would be graduating that year, like me, but her post-graduation plans were set. She'd be going to the United States for college, most likely to study mathematics under her aunt, Dr. Narabikura. Not that she had personally confirmed the latter part of that, but I was pretty sure I wasn't far off the mark. In any case, my time in high school with Miruka would soon come to an end—a fact I knew would occur with one-hundred percent certainty. She would be heading overseas, and I'd be left behind in Japan.

"Miruka's so cool," Yuri sighed. "You must feel like you're being abandoned."

"Why would I?" I snapped. Not that I meant to, but Yuri's seemingly reading my mind triggered something in me.

Shrugging off my outburst, Yuri turned the book she'd been looking at toward me. It was opened to a page showing a map.

"Why do airplanes take these curvy routes?" she asked. "Seems like they should just fly straight to their destination."

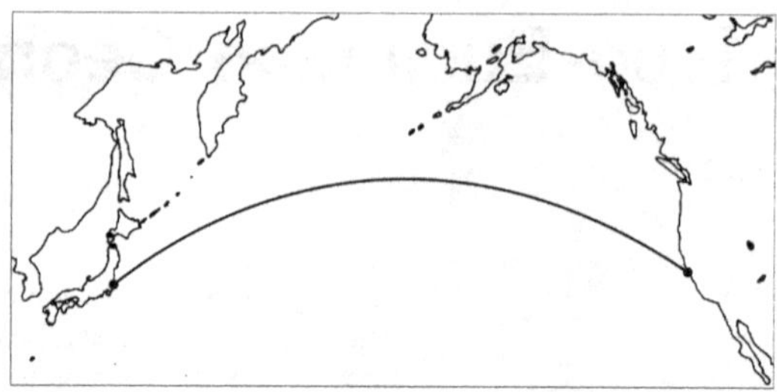

"Ah, right," I said, trying my best to improve my tone. "That's just because of how the map is made. The plane is actually flying in a straight line, but the curvature of the earth makes it look like an arc."

Yuri gave her chestnut ponytail a shake. "Flying in a straight line makes a curve? That doesn't make sense."

"Mmm, maybe instead of 'straight line' I should have said 'shortest path.' The shortest path on a two-dimensional map isn't necessarily the shortest path on a three-dimensional globe. You've studied latitude and longitude in school, right? Those are both circles on a globe, but latitudinal circles are largest at the equator, and shrink as they get closer to the north and south poles. Longitudinal circles are all the same size."

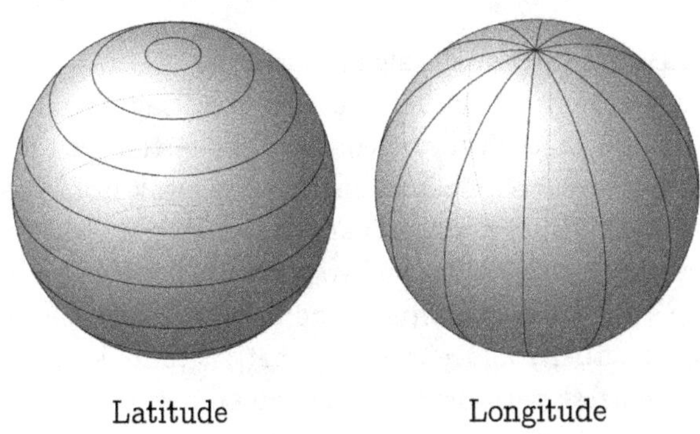

Latitude Longitude

"On a map drawn using the Mercator projection, the latitudes are horizontal lines and the longitudes are vertical lines, but each have lines of the same length."

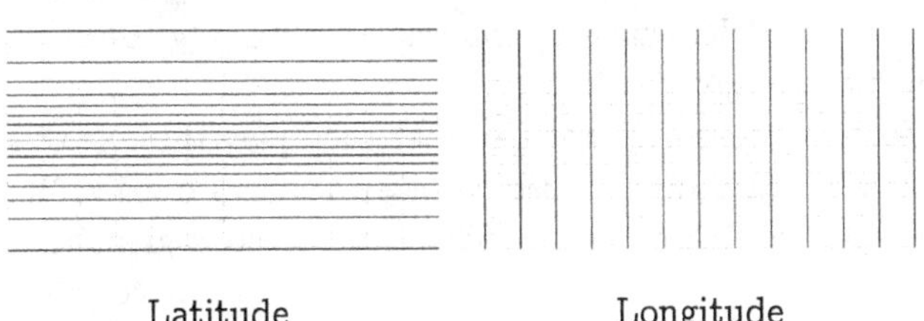

Latitude Longitude

"Okay, so . . . ?"

"Well, that means the closer you get to the poles, the more the latitudinal lines are being stretched out beyond their actual size. In other words, moving a given distance on the map, the closer you are to the equator, the farther you've actually moved."

"Interesting! And that ties into the airplanes thing?"

"Sure. If an airplane were to fly east at a fixed latitude, then on a map it would be heading straight to the right, right?"

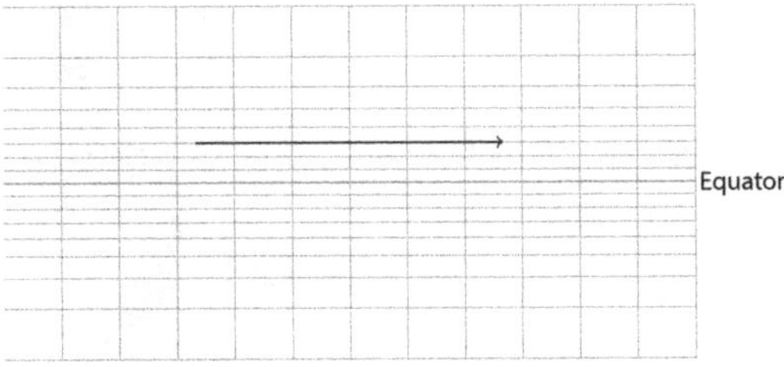

"Sure. But that's not the shortest course?"

"Not necessarily. Because if you've moved, say, an inch *on your map*, the closer you are to the equator, the farther you've actually moved *over the earth*. That means moving away from the equator can shorten the distance you actually travel. A good way to see that is by stretching a string between two points on a globe. If both points are on the same circle of latitude, you'll see that the middle part of the string is actually above or below that latitude. Mathematically, that's the shortest path between the two points."

"So that's the path an airplane should take?"

"Ignoring other considerations like weather or the jet stream, but yeah. You can think of taking a slice of the earth through the plane formed by three points—the point you leave, the point you want to arrive at, and the center of the earth. The shortest path will be along the resulting circle, which by the way is called a 'great circle.'"

"How presumptuous."

"Well, they're pretty great in that they play the role of straight lines on a globe. You can also consider the arcs created by partial great circles as line segments. The most obvious great circle on a globe is the equator, which is the only circle of latitude that is also a great circle. So if you've chosen two points on the equator, the shortest path between them will stick to the same latitude. All circles of longitude are great circles."

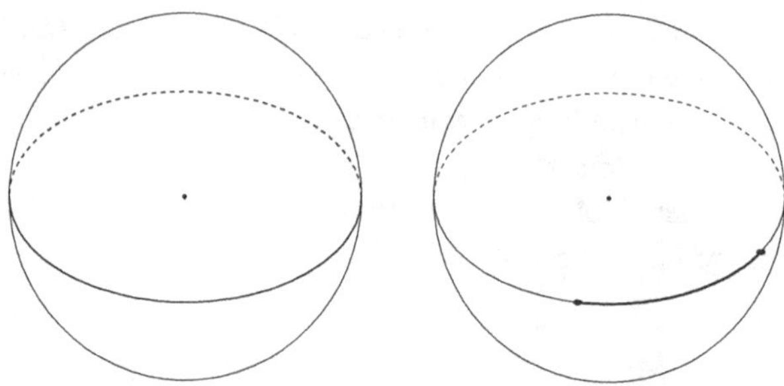

The equator is a great circle. Arc of a great circle.

"I get it," Yuri said, "but this feels weird."

"How so?"

"Because straight lines are supposed to be straight, *always* straight, and stretchable as far as you want! But if you use these great circles as lines, they curve around and run into themselves!"

"Very nice. You've noticed that what we're calling 'straight lines' on a sphere don't stretch out to infinity, they've lost that property. But we're only considering great circles as straight lines in the sense of the curve passing through the shortest path. So really it would be more precise to call these geodesics, rather than straight lines."

"I should have known a new word would pop up."

"Makes sense though, right? Because spheres and planes are very different. They also have some very interesting properties. For example, two straight lines in a plane can never intersect at exactly two different points."

Yuri cocked her head. "Two lines in a plane? Well of course they only intersect at one point. Er, no... They can intersect at one point, no points, or all points!"

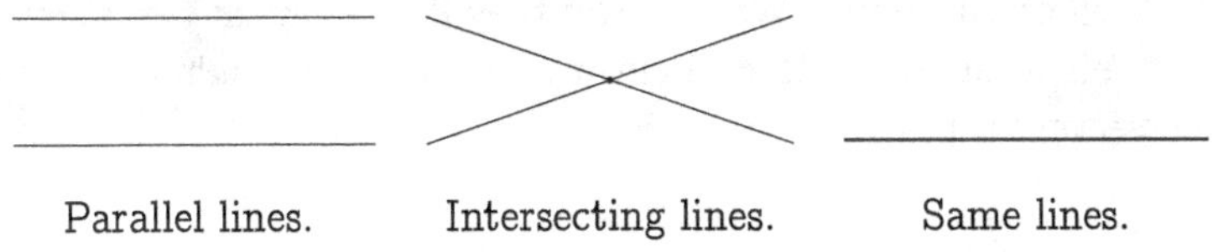

Parallel lines. Intersecting lines. Same lines.

"That's right. Those are properties of straight lines in a plane. But let's see what happens on a sphere, using great circles as our 'lines.' If you have two great circles, they—"

"Hold up! Let me think... Ah, got it. They can only either intersect at two points, or be the same great circle!"

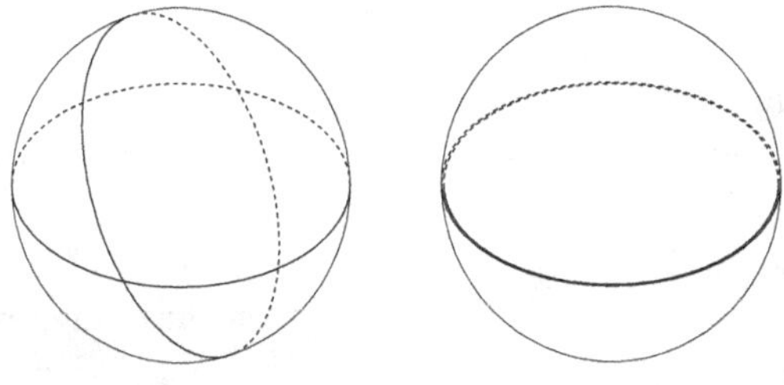

Intersecting great circles. Same great circles.

"You got it. And this means there's no such thing as parallel lines on a globe."

"How so?"

"Well, in a plane if you have a line l and some point P not on l, there's only one line through P that doesn't intersect l, right?"

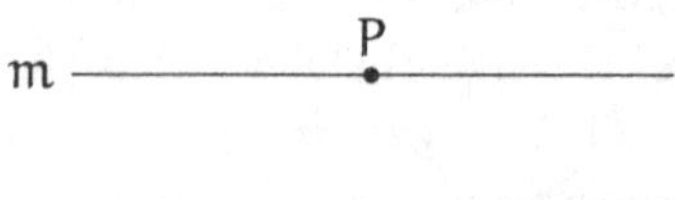

Only one nonintersecting line passes through a point P not on line l.

"I'll just stick with 'they don't touch,' but go on."

"Okay, so what happens on a sphere? If you have a point P that isn't on a great circle l, can you draw another great circle that doesn't intersect l?"

"Well of course not. We just said they'll intersect at two points."

"Well what about this circle m? It passes through P without intersecting l, right?"

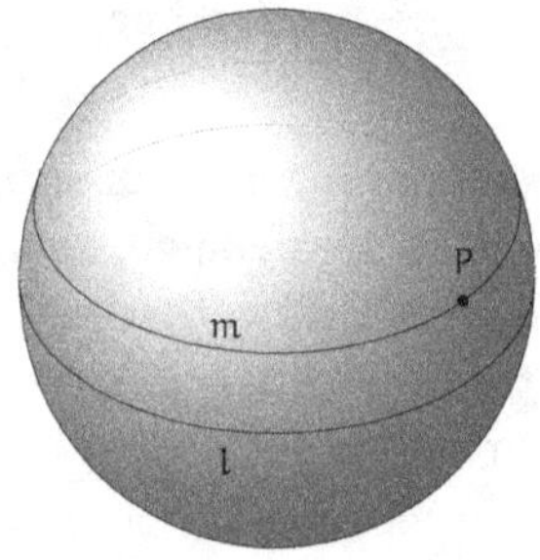

"But that's cheating, because m isn't a great circle! The center of m isn't the center of the sphere, so m isn't what we're calling a straight line!"

"Very good. You've got a good grip on all this."

"So what else ya got? You said spheres have interesting properties, plural."

"Getting into this, aren't you? Let's see... Oh, I know. In a plane, the sum of angles in a triangle is always 180°, right? But the angles in a triangle on a sphere will always add up to *more than* 180°."

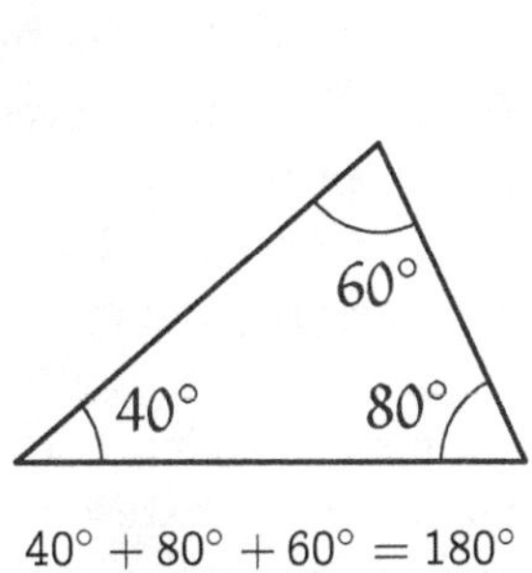

$40° + 80° + 60° = 180°$

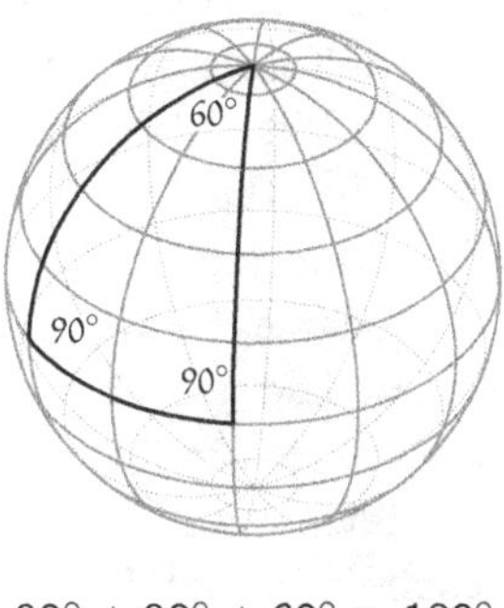

$90° + 90° + 60° > 180°$

"Ah, right. Because triangles kinda like, swell up."

"That's right. So on any given sphere, the larger the triangle, the larger the sum of its angles."

"Hmmm... Hey, if you have three angles, they uniquely describe a triangle, right?"

"That's right, they do. And that holds on a sphere, too. If you use three great circles to create three angles, you've created a triangle on the sphere. There's a difference here, though, in that you can change the size of a triangle in a plane without changing its shape, but that's not true on a sphere. In other words, congruence and similarity are the same thing in spherical geometry."

"Okay, I'm totally using that, next time I see *him*!"

4.2 Between Now and the Future

4.2.1 At School

On Monday, Tetra and I ate lunch together on the roof. I described to her what Yuri and I had talked about, regarding triangles on planes and spheres.

"So in spherical geometry, congruence and similarity are the same?" Tetra said. "That's kinda' neat! Also, how you can consider circles on spheres as lines, and how lines aren't infinitely long."

I nodded. "I agree, it is pretty cool."

"I'm so jealous of Yuri! She gets to talk about fun stuff like that with you all the time."

"I'm just telling her about things I've read in books. It's not like it's stuff I discovered on my own."

"But it's still important to learn new things," Tetra insisted, "whether they come from books, or from your own discoveries, or from things other people teach you, or wherever!"

"I agree. By the way, kinda chilly up here, isn't it?" The rooftop breeze had gone beyond refreshing; with the progressing change in seasons, it now had a bite to it.

Tetra didn't seem to hear me. "I've been thinking a lot about cooperation lately," she said. "About how we all have our strong points, and how by working together we can accomplish things we could never do alone. But also about how the me of the future will be able to do things I can't do now. I want to learn through my relations with others."

"That's interesting, Tetra, and I want to hear more, but . . . maybe inside? It's getting cold. Can we head back in?"

"Spring breezes and autumn winds are different, aren't they?" she said, packing up her bento box. "Spring breezes carry joy, but there's something lonely about the wind in fall."

"Indeed. But this is starting to feel more like a winter wind. Maybe today should be the last time we meet up here for lunch."

"I'm sad to say I must agree."

"I wonder if this will end up being my last rooftop lunch as a high school student?" I mused. "By the time the weather warms up enough next spring, I'll have graduated."

"Oh, I . . . I hadn't thought of that . . . "

Tetra looked away, and the afternoon bell rang.

4.3 Hyperbolic Geometry

4.3.1 Learning

My discussion with Tetra left me all the more conscious of how short my remaining high school career was. There were so many things I could tack "for my last time in high school" onto. Just a few more months until college entrance exams, then I'd be graduating, regardless of the results. Most of my time until then would be filled with studying. Not that I hated studying, and not to say I wasn't aware of how important college would be to my future. There was still plenty of learning on whatever path might lay before me. New people, too.

New people? I figured Tetra's talk of learning through her relations with others had struck a chord. Indeed, much of what I'd learned during my time in high school was thanks to having met her, and Miruka, and Mr. Muraki, and others. Surely the same would happen in college, which would bring even more opportunities for meeting new people and learning new things.

Gotta get there first, though.

4.3.2 Non-Euclidean Geometry

After classes I headed to the school library, where I found Miruka and Tetra already deep in discussion. I slid into a seat just in time to catch the start of Miruka's lecture.

"—but the first stepping stone to *non*-Euclidean geometry is of course Euclidean geometry, the kind we've been learning at school. It got its start around 300 BC, with the thirteen volumes of Euclid's *Elements*. That work isn't exclusively about geometry, but it's remarkable in that its style of providing definitions and postulates, then presenting proofs based on them, became the model for how we make mathematical assertions even today. Here are some definitions from *Elements*."

1. A point is that which has no part.
2. A line is breadthless length.
3. The ends of a line are points.
4. A straight line is a line which lies evenly with the points on itself.
5. A surface is that which has length and breadth only.
6. The edges of a surface are lines.
7. A plane surface is a surface which lies evenly with the straight lines on itself.

⋮

23. Parallel straight lines are straight lines which, being in the same plane and being produced indefinitely in both directions, do not meet one another in either direction.

"Saying 'A point is that which has no part' doesn't look like a normal definition, so it's better to think of this as a declaration that we'll be using the word 'point' as a technical term. Particularly important are the postulates that followed Euclid's definitions. These are declarations of some facts and abilities that we will take for granted, without proof. In other words, postulates are like propositions we can use without having to prove them. You can think of postulates as stating what we're allowed to do using points and straight lines, or as how we determine the meaning of technical terms. Postulates are also called axioms. Here's how Euclid set up his postulates."

Let the following be postulated:

1. To draw a straight line from any point to any point.
2. To produce a finite straight line continuously in a straight line.
3. To describe a circle with any center and radius.
4. That all right angles equal one another.
5. That, if a straight line falling on two straight lines makes the interior angles on the same side less than two right angles, the two straight lines, if produced indefinitely, meet on that side on which are the angles less than the two right angles.

"Axioms are very important in Euclid's *Elements*, in that it uses them to prove all its theorems, but there's a big problem with what we've read so far..."

Miruka fell silent.

"The problem being...?" Tetra asked.

"The parallel postulate?" I suggested. I knew enough of non-Euclidean geometry to know that the parallel postulate always came up.

"That's right," Miruka said. "Can you read Euclid's fifth postulate for us, Tetra?"

She did so.

> That, if a straight line falling on two straight lines makes the interior angles on the same side less than two right angles, the two straight lines, if produced indefinitely, meet on that side on which are the angles less than the two right angles.

"Long, isn't it!" Tetra said. "So this is the parallel postulate?"

"It's easier to understand with a figure," I said. "Start with a straight line intersecting two others. Here's a line n intersecting lines l and m."

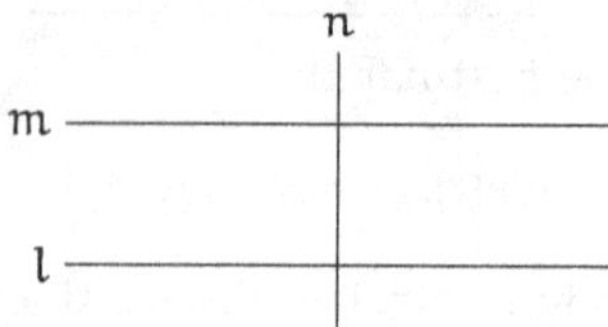

"Now we make the sum of the interior angles on the same side smaller than two right angles."

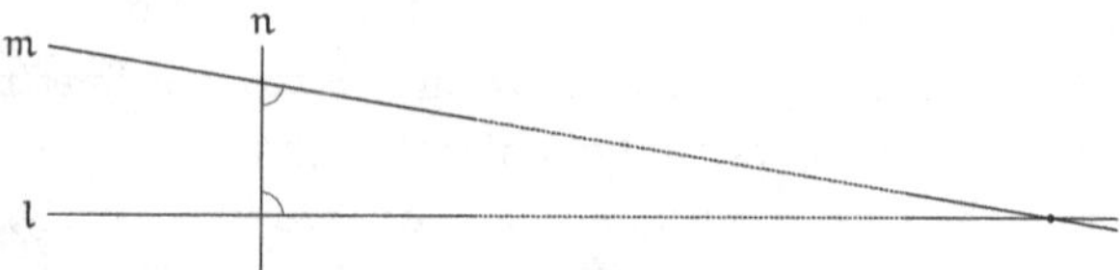

"See? When we do that, the lines intersect on the side with the angle less than a right angle."

"Okay, sure. I see it now," Tetra said. "Well, except for the part about how this is a problem."

Miruka tapped my illustration. "This is simply a description of what the parallel postulate says. The question is, do we really need to establish the parallel postulate as axiomatic?"

"Because it's too long to be an axiom?"

"It is very long, yes, especially when you compare it with the other four. But its length isn't the problem. What bothered mathematicians over the years was a nagging doubt that it could be proved *using only the first four postulates.* Euclid took his postulates as prerequisites for the proofs he came up with. If the parallel postulate could be proven from the others, there would be no need to elevate its assertion to axiomatic status, and we'd prefer to work from the fewest base assumptions possible. Euclid apparently thought it was appropriate as an axiom, but what if he was wrong?"

"I guess that *would* be a problem . . . "

"Many mathematicians took up that challenge. What better way to leave your mark, after all, than by proving Euclid wrong? But none were successful."

"So it's an unsolved problem?"

"Not quite. In the eighteenth century, a mathematician named Giovanni Saccheri attempted to demonstrate the parallel postulate through a proof by contradiction. In other words, he assumed this axiom to be incorrect, and tried to show that doing so would result in a contradiction. Saccheri's research produced some nonintuitive, odd results, which he thought were contradictory and thus indicated a successful proof. What he'd shown wasn't logically inconsistent, though. Actually, we can consider his discoveries to be somewhat prophetic."

"Prophetic? Yeah?" I said. I knew the basics of non-Euclidean geometry, but I'd never heard of this Saccheri, or any prophetic discoveries he might have made.

"A discovery like this," Miruka said, a gleam in her eye.

Assuming the parallel postulate does not hold, two "straight lines" in a "plane" will either

- diverge endlessly at both ends, or
- diverge endlessly at one end and converge endlessly at one end.

"It's like some strange puzzle," Tetra whispered.

"Saccheri also showed that if we create plane geometry without assuming Euclid's parallel postulate, then the sum of interior angles in a triangle would be less than 180°. But again, this isn't necessarily a logical contradiction. Later, Lambert showed that we can derive the parallel postulate if there exists two triangles that are similar but not congruent, but that still wasn't a proof of the parallel postulate itself."

"Triangles that are similar but not congruent? I guess you couldn't create that on a sphere," I said.

"In any case," Miruka continued, "no one was able to prove the parallel postulate holds, and no one could prove that it doesn't. Finally, Bolyai and Lobachevsky discovered non-Euclidean geometry in the nineteenth century."

"Oh, how interesting!" Tetra said. "They weren't able to do it on their own, so they worked together on a proof?"

"No. They made their discoveries independently, unaware of each other's research. Even so, non-Euclidean geometry arose in different places at approximately the same time. The great mathematician Gauss may have discovered it even before Bolyai and Lobachevsky."

"So what happened? Did they prove that the parallel postulate holds, or that it doesn't?"

"Neither one."

"Wait, what?"

4.3.3 Bolyai and Lobachevsky

Miruka continued her talk.

"So Saccheri looked for contradictions that would arise from assuming the parallel postulate does not hold. By contrast, Bolyai and Lobachevsky considered how they could use axioms other than the parallel postulate to systematize geometries differing from Euclidean geometry. Thus, non-Euclidean geometry. Euclidean geometry is a system originating from five postulates, including the parallel postulate. Bolyai and Lobachevsky removed that postulate to create a new geometry starting from a different set of five postulates. Today, what they came up with is called hyperbolic geometry. Using numbers of straight lines, it looks something like this."

- Spherical geometry
  There exists no line passing through a point P not on line l that does not intersect l.

- Euclidean geometry
  There exists one line passing through a point P not on line l that does not intersect l.

- Hyperbolic geometry
  There are at least two lines passing through a point P not on line l that do not intersect l.

"I have a question," Tetra said, raising her hand. "So there's spherical geometry, Euclidean geometry, and hyperbolic geometry, but... which one is the *real* geometry?"

"No one of these is the 'real' geometry," I said. "Or you could say they're *all* real geometries. A non-Euclidean geometry is just as valid as Euclidean geometry. Math is all about setting up some axioms and seeing what we can prove, what mathematical assertions we can make by using them. So they're all correct, they're just based on different axioms."

Tetra chewed on a nail for a moment while considering this. After a short time she brightened.

"This is another example of the I-know-nothing game! Why didn't I see that, considering how many times we've done the same thing? We did the same thing when we talked about groups[1], and when we talked about number theory[2], and when we talked about probability[3], and when we talked about topological spaces... In every case, we just looked at a set of axioms to see what pops out. So I guess geometry is the same?"

"It is," I said.

"Mathematicians propose axioms," Miruka said, "and establish theorems based on what they can prove using those axioms. This is why we can study mathematics without being constricted by the real world. Not to say that's necessarily what was in Euclid's mind..."

Tetra twisted her face. "I'd always thought of geometry as being about everyday shapes. But now you're saying it isn't related to the real world?"

"I didn't say 'unrelated.' Historically speaking, geometry likely arose from a desire to better understand the world around us. But mathematics does *not* say anything about what kind of geometry our reality, the universe we live in, is based on."

Tetra sank back into silence, and Miruka and I gave her some space.

I considered how we were sitting in a library, which was in a school, which was in a town, which was in a country, which was on the earth, which was floating in space. A tiny planet, home to a tiny country, containing a tiny town and a tiny school and a tiny library, in which the three of us sat thinking. Yet what we were thinking about was shapes that transcend space.

---

[1]See *Math Girls 2: Fermat's Last Theorem.*

[2]See *Math Girls 3: Gödel's Incompleteness Theorems.*

[3]See *Math Girls 4: Randomized Algorithms.*

Miruka suddenly stood from her chair. She looked up, triggering a series of waves in her long black hair.

"What do we really know?" she said. "We know about lines, and we know about parallel lines. We think we do, at least. But do we really? We can picture lines in our imagination. Same with parallel lines. When told to imagine a line and a point not lying on it and another line passing through that point, we can piece together a mental image, despite the many possibilities for that situation and the presence of shapes stretching off to infinity. When asked if such a situation is possible, we want to say yes. We want to affirm the uniqueness of a parallel line passing through that point. So long as our lines don't bend anywhere, we say, a single parallel definitely exists.

"We feel like we're able to say so with certainty, so why consider a geometry where parallel lines don't exist? Why dream up geometries where more than one parallel line can exist? Is it because mathematicians are fantasists, untethered from reality? No. Because they're agnostics, claiming we cannot know what lies at infinity? No. Because they're pessimists, insisting intuition is always faulty? No. Because they're realists, aware that we can never actually draw lines that are parallel in the truest sense? Again, no.

"Mathematicians simply trust in logic. By replacing the parallel postulate with some other axiom, a different geometry is born. What a fantastic idea.

"It took some time for the world to accept non-Euclidean geometry, similar to how Galileo started to doubt his own discoveries.[4] Which is understandable, since you don't have to be a Galileo to feel uneasy when learning you can create a one-to-one correspondence between natural numbers and squares. Even so, infinity can indeed be defined as a correspondence between a whole and its parts. The parallel line axiom is the same. If we can't prove that axiom, maybe we can just use the propositions of other axioms to create a *different* geometry. A complete reversal of thinking. The demand for uniqueness of parallel lines creates Euclidean geometry, and insisting on their plurality creates something else."

---

[4]See *Math Girls 3: Gödel's Incompleteness Theorems*, chapter 3.

With that, Miruka plopped into the chair beside me.

"Think of a geometry as a structure built from what you can derive from its axioms," she said, leaning in close to my face.

"I... I do!" I protested, reflexively pulling back.

"A question!" Tetra said, her outstretched hand coming between us. "I know Euclidean geometry is the 'normal' geometry we do on planes, and I can imagine spherical geometry by just thinking about drawing shapes on a globe. But hyperbolic geometry? What is that, specifically? I'm not sure how I'm supposed to picture it in my head."

"You're not alone," Miruka said. "One reason why people had such difficulty accepting non-Euclidean geometries is because they were so convinced of the self-evident truth of the Euclidean parallel postulate. It was only after mathematicians like Klein and Poincaré and Beltrami constructed some non-Euclidean models that non-Euclidean geometries gradually came to be accepted."

"Models? What do you mean by—"

"The library is *closed*!"

The sudden announcement caused each of us to jump. *Already?*, I thought, but I glanced out the window and saw that sure enough, it was already getting dark. Once again, I'd become so entranced by Miruka's lecture I'd lost all sense of time.

4.3.4 At Home

After eating dinner with my mother that night, she moved to the kitchen and started washing.

"Dad's going to be late again?" I asked, collecting plates from the table.

"I imagine so," she said, lathering up a sponge.

Now that Mom was recovered and back home, Dad had returned to his usual schedule of daily overtime.

"Sorry to have caused so much trouble," Mom said. "Just when you need to be focusing on your studies."

"Nah, no big deal."

"I'm sure I caused all kinds of problems for your dad at work, too. And the money we had to spend..."

"That much?"

"Well, nothing for you to worry about." She suddenly brightened. "How's *your* work coming, by the way?"

"My work?"

"Well, studying for exams is your job now, right?"

"Ah. It's going okay, I guess."

"I think I'll do the washing later," she said. "Do you want a cup of cocoa? I also have some rooibos tea..."

"I'm good. I might make a cup of coffee later."

Mom tsked. "You should avoid caffeine at night, you know."

I shrugged and headed up to my room.

Not everything was back to how it was, I realized. Since Mom had come home, I'd started to notice how frequently she seemed to be tired, and the wrinkles that had appeared at the corners of her eyes. Or maybe she'd been like that from before, but now my perspective had changed. In any case, I was now keenly aware that my mother was aging. Another sign of time marching cruelly on.

A couple of hours later, between practice problems I'd been working on, I recalled Tetra and her talk of wanting to cooperate with others. How working with others allowed us to accomplish things we couldn't do on our own. My college entrance exams, though? That was all up to me. I'd have to solve those problems on my own, applying my own abilities. Because that was "my job" now.

I supposed the same held for mathematical discoveries, or proofs of theorems. As Miruka had told us just that day, while the development of non-Euclidean geometries is attributed to people like Bolyai and Lobachevsky and Gauss, they each made their own independent discoveries.

I'd told my mother I was doing okay at "my job," but was I really? Only peering into the future could tell me that. Not even very far into the future, just to next spring. Actually, just seeing a single point in time—that moment where I got back my test results—would be enough.

Ah well, back to it...

4.4 Setting Aside the Pythagorean Theorem

4.4.1 Lisa

Leaving my classroom after my last class the next day, I found a red-headed girl waiting outside the door. She was clutching a notebook computer, and had a hairstyle that looked like it was coiffed using garden shears.

"Hi, Lisa," I said. "What's up?"

"You are summoned," she rasped in her husky voice.

"Summoned? By who?"

"Miruka," she replied, already heading down the hallway.

Lisa was a first-year student, and Miruka's cousin. She was a very talented programmer, and I never saw her without her computer, which was the same brilliant red as her hair. I was surprised when she led me not to the library, but the audiovisual room. A screen was pulled down in front of the large blackboard there.

"There you are," Miruka said. She was sitting on the podium, her long legs crossed.

"Glad you could make it!" Tetra said from her seat in the front row.

"What's going on?" I asked. "I was planning on going to the library."

"We have some unfinished business from yesterday," Miruka said. "Namely, models of non-Euclidean geometries. Won't take long, but I think you'll be interested."

As Miruka spoke, Lisa was connecting her computer to the AV equipment in the room and making some adjustments.

"Going dark," she warned, then tapped a key. Heavy curtains closed over the windows, and the ceiling lights dimmed. The projector came on and lit up the screen.

4.4.2 Defining Distance

"Given points P and Q in the plane," Miruka said from somewhere in the darkness, "the shortest path between them is a straight line, like the line segment $\overline{\mathrm{PQ}}$ here."

A figure appeared on the screen.

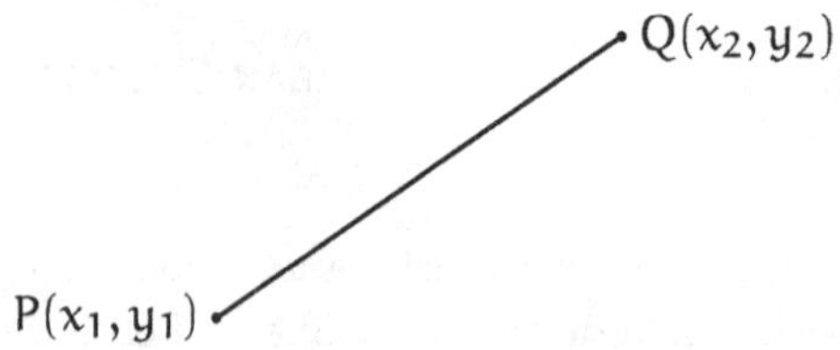

My eyes started adjusting to the dark, and I realized I could see everyone's faces, dimly illuminated by the screen.

"Of course," Miruka continued, "we can only talk about shortest paths if we have a definition for the distance between two points. Without that, we couldn't say whether a path is long or short. Okay then, so how *do* we define the distance between two points? Tetra?"

"We calculate it from the coordinates of those points," Tetra said. "If we have points $P(x_1, y_1)$ and $Q(x_2, y_2)$, we can calculate the distance as $\sqrt{(x_2 - x_1)^2 + (y_2 - y_1)^2}$."

"Very good. And behind that calculation lies the Pythagorean theorem."

Lisa tapped a button and the slide on the screen changed.

$$\text{Distance}^2 = (x_2 - x_1)^2 + (y_2 - y_1)^2$$

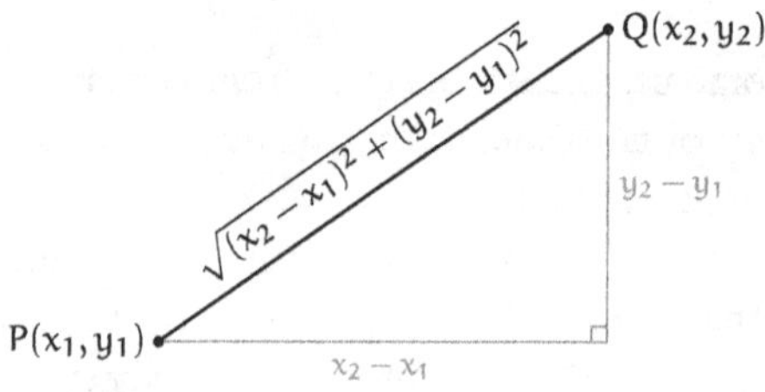

"Here we've defined distance according to changes in x and y coordinates. Things will be the same no matter how minuscule those differences are, so let's use dx to mean an infinitesimal change in a x coordinate, and dy to show the same for a y coordinate. Then we can define ds as an infinitesimal change in distance. It's also called a line element. The relation between dx, dy, and ds looks like this."

$$ds = \sqrt{dx^2 + dy^2}$$

"We can therefore also represent it like this."

$$ds^2 = dx^2 + dy^2$$

"So Euclidean geometry uses the Pythagorean theorem to define distance, or we can conversely consider a geometry using the Pythagorean theorem to define distance as Euclidean geometry. But let's set the Pythagorean theorem aside for a bit, and see if there isn't some other way we can define distance."

"Set it aside?" I said. "We can... do that?"

"Sure. And by defining a new distance, we can develop a new geometry within Euclidean geometry. In other words, we'll create a model on top of Euclidean geometry. When we use the Pythagorean theorem to define distance, points on a shortest path will form a straight line. But if our definition of distance changes, shortest paths may not appear to our eyes as being straight. This is the same as how while great circles describe shortest paths on a globe, they can look like curves on a map. So let's try creating the hyperbolic geometry of Bolyai and Lobachevsky as a model on top of Euclidean geometry.

"We'll start with the Poincaré disk model."

4.4.3 The Poincaré Disk Model

A large circle appeared on the screen.

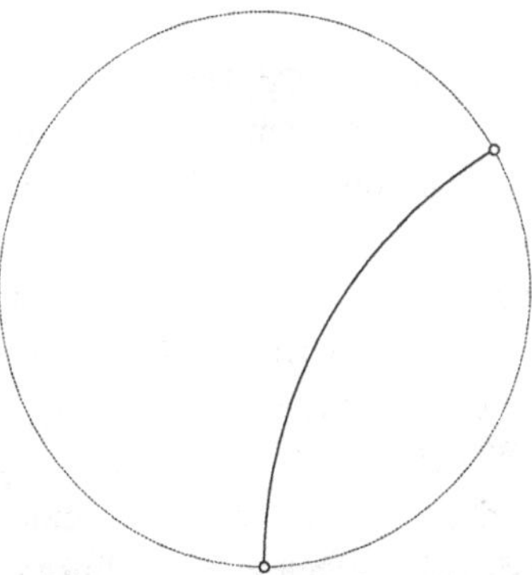

The Poincaré disk model.

"This is the the Poincaré disk model?" Tetra asked.

"It is," Miruka replied. "This shows a single 'straight line' drawn on a 'plane.' In this model, the plane is the interior of a circle with radius 1, centered on the origin in planar coordinates. The periphery of this circle is not included in this plane. In other words, the plane in the Poincaré disk model is the region D, which we describe like this."

$$D = \{(x, y) \mid x^2 + y^2 < 1\}$$

"How strange. The entire plane, captured inside a circle..."

"Which also implies that points in this model are those points within the disk. Straight lines are circular arcs that are orthogonal to the disk's boundary. As a special case, line segments describing the disk's diameter are also considered as straight lines. Neither kind of line, however, includes the points at which they would intersect with the disk's boundary."

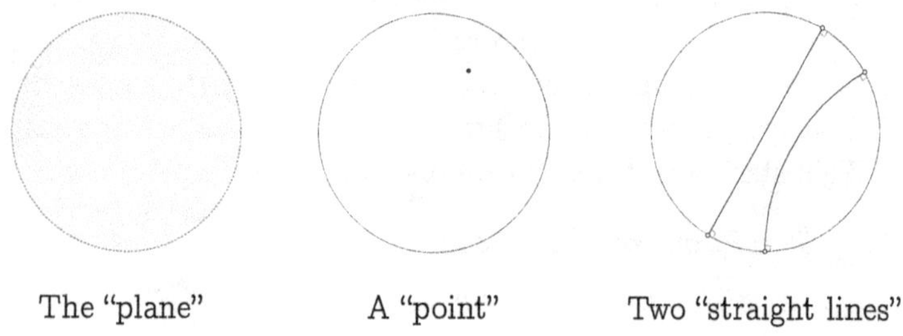

The "plane" A "point" Two "straight lines"

"The figure on the right here shows two straight lines in the Poincaré disk model. One of them is an arc, so it doesn't look like a shortest path. It is, though, according to how the Poincaré disk model defines distance."

"And it looks curved because a 'straight line' here is the arc of another circle, which is determined by orthogonality with the circumference of the disk!" Tetra said.

Miruka nodded. "That's right. So let's see what parallel lines look like in the Poincaré disk model. You have some line l, and point P not on that line. So drawing other straight lines through P that don't intersect l, we get— Lisa?"

Lisa tapped a key to move to the next slide.

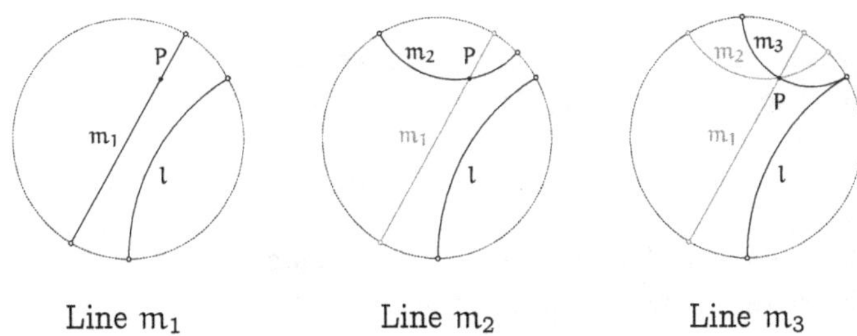

Line m_1 Line m_2 Line m_3

"These three lines m_1, m_2, m_3 pass through P, but don't share any points with line l. I'm only showing three lines here, but of course there are infinitely many possibilities. So in Euclidean geometry only one parallel line is possible, but in hyperbolic geometry there are infinitely many parallel lines."

"Hold up," Tetra said. "Isn't m_3 here touching l? It sure looks like they share a point at the end there..."

"But don't forget what a line is on a Poincaré disk," I said. "Miruka said points on the periphery aren't included on the line. The plane in a Poincaré disk is only the disk's interior, so the periphery isn't even on the plane. So I don't think l and m_3 share a point. Right?" I glanced at Miruka.

She nodded. "That's right. Do you see how the Poincaré disk model is an excellent example of Saccheri's prophecy? Specifically, the relation between l and m_1, or between l and m_2, is that both diverge without limit, while the relation between l and m_3 is diverging without limit on one end while approaching without limit on the other."

"I'm sorry, something I don't quite get about that," Tetra said. "You say they're diverging without limit, but it's not like the lines are extending forever outside of the disk, right? It's been bothering me since the beginning, that what we're calling 'lines' in a Poincaré disk don't extend out to infinity. So aren't lines in hyperbolic geometry finite, kind of like they are in spherical geometry?"

"No, they aren't. Lines in the Poincaré disk model are just as infinite as they are in Euclidean geometry, because we're defining 'distance' differently here."

Tetra cocked her head. "We've defined a different distance?"

"Recall how in a Euclidean plane we can represent the line element ds according to the Pythagorean theorem."

$$ds^2 = dx^2 + dy^2$$

"A line element ds in the hyperbolic plane of the Poincaré disk model is different, though, like this."

$$ds^2 = \frac{4}{(1-(x^2+y^2))^2}(dx^2 + dy^2)$$

"Comparing the two, we see that we're multiplying by this factor."

$$\frac{4}{(1-(x^2+y^2))^2}$$

"This is the factor by which we're deviating from the Pythagorean theorem. Generally, a function used to determine distances in a space is called a metric. By defining our line element, we add a metric to the space, allowing us to calculate distances there."

Line elements in a coordinate plane model of Euclidean geometry
$$ds^2 = dx^2 + dy^2$$

Line elements in the Poincaré disk model of hyperbolic geometry
$$ds^2 = \frac{4}{(1-(x^2+y^2))^2}(dx^2 + dy^2)$$

I took a moment to compare the two. Something dawned on me.

"So the closer the point is to the disk's periphery, the larger ds is, right?" I said. "Because moving closer to the periphery makes the $1-(x^2+y^2)$ in the denominator smaller."

"Well spotted," Miruka said. "Yes, the line element ds changes according to its Euclidean distance $\sqrt{x^2+y^2}$ from the origin. Pretend there's a point moving at constant speed in the Poincaré disk model. Observing that point, we would see it seeming to slow down as it approached the edge of the disk, to the extent that it would never reach the edge, no matter how long we watched."

Tetra did a double-take. "Wait, it's moving at constant speed, but getting slower? Isn't that a contradiction?"

"No, because we're talking about constant speed with regards to distance as represented in the Poincaré disk model. It appears to slow down to us, because our observations from the outside are based on distance as measured in Euclidean geometry."

"Oh, right. Okay..."

Miruka paused, tapping her lips. She whispered something to Lisa, who loaded a new set of slides, then resumed her talk.

"Let's use integrals to define the distance that the point moves. We'll let $(x(t), y(t))$ represent the position of our point at time t. As t changes, the point will move, describing a curve. The speed in the x and y directions are respectively $\frac{dx}{dt}$ and $\frac{dy}{dt}$, so the movement speed $\frac{ds}{dt}$ looks like this."

$$\left(\frac{ds}{dt}\right)^2 = \left(\frac{dx}{dt}\right)^2 + \left(\frac{dy}{dt}\right)^2$$

$$\frac{ds}{dt} = \sqrt{\left(\frac{dx}{dt}\right)^2 + \left(\frac{dy}{dt}\right)^2}$$

"In Euclidean geometry, we can integrate $\frac{ds}{dt}$ with respect to t to find the distance moved. Doing so gives us the length of the curve the point describes as t moves from a to b, in other words, the length of the path between points $(x(a), y(a))$ and $(x(b), y(b))$. Specifically, the integral looks like this."

$$\int_a^b \sqrt{\left(\frac{dx}{dt}\right)^2 + \left(\frac{dy}{dt}\right)^2}\, dt$$

"We can do the same thing in the hyperbolic geometry of the Poincaré disk model, except now our curve length looks like this."

$$\int_a^b \frac{2}{1-(x^2+y^2)}\sqrt{\left(\frac{dx}{dt}\right)^2+\left(\frac{dy}{dt}\right)^2}\,dt$$

"As we've said, the line element ds in the Poincaré disk model becomes larger as we approach the disk's edge. So a given length as perceived by Euclidean eyes will appear longer to hyperbolic eyes when it's closer to the periphery. This is the same concept as on a map with a Mercator projection, where distances increase as you approach the poles.

"So Tetra, before you said there are no infinite lines in the Poincaré disk model, and that's true if you're thinking of lengths in Euclidean terms. But that's not the case when we use this model's concept of 'length.' A hyperbolic resident of this plane would perceive the disk's periphery as an infinitely distant horizon, unreachable no matter how long she travelled toward it. *That* is the nature of the Poincaré disk model.

"Tilings probably give a clearer way of showing this. We can tile the Euclidean plane using triangles, squares, or hexagons like this."

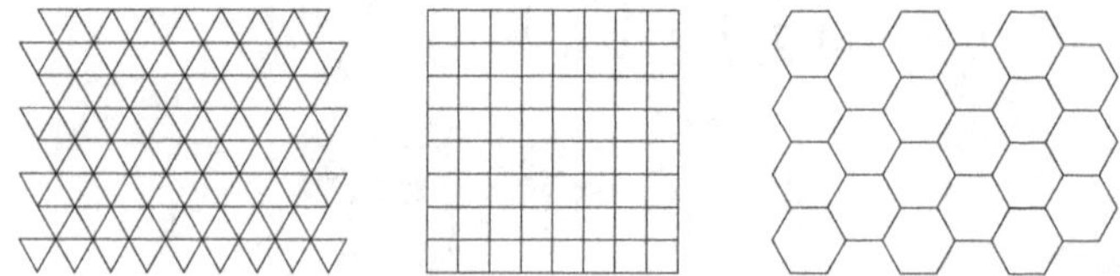

Tiling the Euclidean plane with regular n-gons.

"We can similarly tile the Poincaré disk using regular polygons as they would appear there."

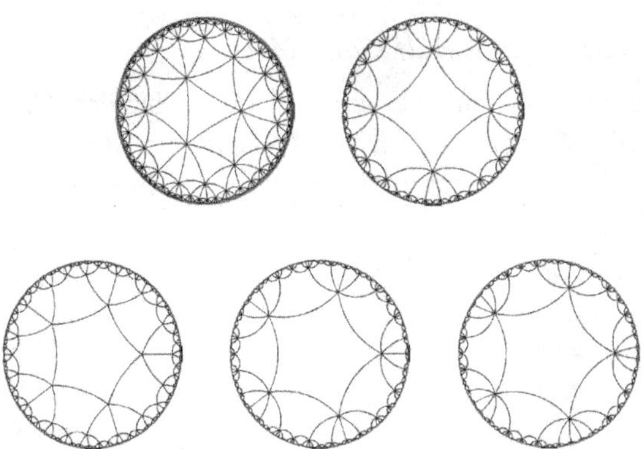

Tiling the Poincaré disk with regular n-gons.

"Ooh!" Tetra squealed. "These remind me of Escher's pictures!"

"Indeed. M.C. Escher created many artworks inspired by the hyperbolic geometry of the Poincaré disk."

"So that's where they come from . . ." I muttered.

"As Saccheri discovered, the sum of interior angles of a triangle in a hyperbolic geometry will be less than 180°. Even so, each side length in a regular polygon will have equal length according to the metric for a Poincaré disk."

"It looks like those sides are getting really tiny when they get close to the edge," Tetra said.

"That's right. Equal-length line elements look longer near the center of the Poincaré disk, and smaller as they move away from it. That's because the metric we're using here determines length according to the Euclidean distance from the disk's center."

"So when we're looking at a Poincaré disk, it's like we're able to see all the way to infinity!"

"Interesting," I said. "I guess the same thing would happen in principle if we could stand on a Euclidean plane and see all the way to the horizon, but this is much more . . . compact."

Tetra turned toward Lisa. "Great job, making all these illustrations!"

Lisa shrugged and mumbled, "Graphics libraries."

"Well I'm looking forward to more!"

4.4.4 The Poincaré Half-plane Model

"The Poincaré disk isn't the only model of hyperbolic geometry," Miruka said. "We can also create a half-plane model, where drawing two 'straight' lines like we did in the Poincaré disk model would look like this."

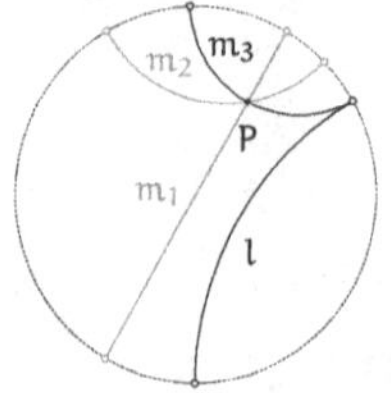

Poincaré disk model

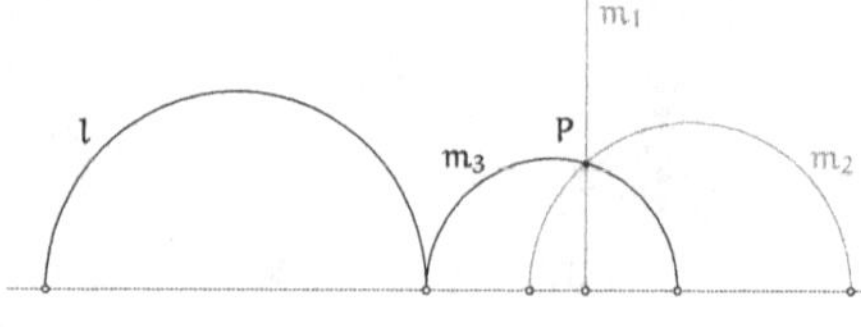

Poincaré half-plane model

"The half plane H^+ is defined like this."

$$H^+ = \{(x, y) \mid y > 0\}$$

"And the line element ds in the half-plane model is defined like this."

Line element in the hyperbolic half-plane model

$$ds^2 = \frac{1}{y^2}(dx^2 + dy^2)$$

"As you can see from this equation, ds is dependent on y. The closer y gets to 0, in other words, the closer it gets to the x-axis, the larger the line element ds becomes. So the periphery of the Poincaré disk in that model has now become the x-axis. To residents of this hyperbolic world, the x-axis is infinitely far away, a horizon they can never reach."

"There's something very romantic about that," Tetra said.

4.5 Going Beyond the Parallel Postulate

"Very interesting," I said. "So this line element ds is some infinitesimal distance that's determined according to where we're looking, namely the position of some point (x, y) in the coordinate plane, right? I like how even when using the same hyperbolic geometry, we can create models that look very different. It's a lot like how there are maps made in so many different ways. The Mercator projection, the Mollweide projection, the azimuthal equidistant projection... All different ways of representing the same earth on a map."

Miruka pointed at me. "That's a very important point. We can generalize geometry according to the way we establish our line element, in other words how we decide the metric we'll use."

"We can generalize... *geometry*?"

What on earth... ?

"Well, recall the path by which non-Euclidean geometry arose from Euclidean geometry, and how the parallel postulate was involved, how Lobachevsky and Bolyai created hyperbolic geometry by replacing it with another axiom. One that says there are infinitely many lines passing through some point not on another line, yet not intersecting that line."

Tetra nodded. "Right. We were able to create three geometries, just by changing how parallel lines work—spherical geometry, Euclidean geometry, and now hyperbolic geometry."

"But there's more," Miruka said. "Riemann went a step beyond even worrying about the parallel postulate. He thought of how we might generalize the way in which we select a metric. Rather than think about some specific geometry, he focused on metrics that create new geometries. Change your metric and you've changed your geometry, so by studying metrics, you're studying all geometries. A geometry you've generalized by introducing a metric is called a Riemannian geometry."

"I've heard of Riemannian geometry before," I said, "but I didn't really know what it meant. I'd always thought it was just another kind of non-Euclidean geometry."

"Actually, that term is used with two different meanings. For one, it can refer to a specific non-Euclidean geometry that Riemann studied. The more important meaning, though, is geometries generalized through introduction of a metric. It's a common term used to describe infinitely many geometries."

"So it's something like this?" Tetra said. "Riemannian geometries include Euclidean geometries and non-Euclidean geometries and infinitely many other geometries I've never even heard of? Meaning, Euclidean geometry is just one example of a Riemannian geometry?"

"That's right. Let's represent the line element in a Poincaré disk like this."

$$\begin{aligned} ds^2 &= \underbrace{\frac{4}{(1-(x^2+y^2))^2}}_{g(x,y)} \left(dx^2+dy^2\right) \\ &= g(x,y)\,\left(dx^2+dy^2\right) \end{aligned}$$

"We can respectively write dx^2 and dy^2 as $dxdx$ and $dydy$. Doing so and explicitly writing out $dxdy$ and $dydx$, we can write ds^2 like this."

$$ds^2 = g(x,y)\,dxdx + 0\,dxdy + 0\,dydx + g(x,y)\,dydy$$

"Let's also say the coefficients $g(x,y), 0, 0, g(x,y)$ are $g_{11}, g_{12}, g_{21}, g_{22}$, and let x, y be x_1, x_2.

$$ds^2 = g_{11}\,dx_1dx_1 + g_{12}\,dx_1dx_2 + g_{21}\,dx_2dx_1 + g_{22}\,dx_2dx_2$$

"In other words, we can write ds^2 like this."

$$ds^2 = \sum_{i=1}^{2}\sum_{j=1}^{2} g_{ij}\,dx_i dx_j$$

"Here, g_{ij} is a function showing the extent to which a point deviates from the Euclidean length in a given direction. There are a few conditions on this g_{ij}, but a metric like this is called a Riemannian metric. A Riemannian metric determines the line element ds, and by integrating that we can find the length of a curve in that space. Given a metric and a curve connecting two points, we can define the curve's length using integrals. Based on that length, we can define the distance between the points. In the case of a Euclidean space, the distance between two points is the length of a line segment connecting them.

"A metric is a generalization of distance. In the Euclidean, spherical, and hyperbolic geometries we've talked about, the metric does not change with orientation. We're used to spaces in which distances are the same in every direction, but we can also consider distances that change with orientation or position. Bolyai and Lobachevsky discovered hyperbolic geometry in their attempts to prove the parallel postulate, but hyperbolic geometry is just one example of a non-Euclidean geometry. As we've seen, by defining an appropriate metric, we can create a model of hyperbolic geometry on top of Euclidean geometry, like we did with the Poincaré disk and half-plane models.

"Riemann went a step further, using metrics to create infinitely many geometries. He gave a lecture describing his ideas, which greatly excited his teacher, Gauss. At this time Riemann was twenty-seven years old, while Gauss was seventy-seven. Perhaps Gauss realized what he was hearing was the future of geometry. Systemization of the geometry that started with the parallel postulate was already headed in a direction toward creating new geometries by bringing in different axioms. Now, Riemann was showing the possibility of setting aside all this focus on the parallel postulate, and instead creating infinitely many new geometries through the application of metrics. This was a step toward the study of the spaces themselves.

"In today's mathematics, those spaces are called Riemannian manifolds."

The last bell of the day chimed.

"Time to go home," Lisa said.

4.6 At Home

That night I was sitting at my desk, a cup of warm rooibos tea next to me—placed there by my mother, of course. I was thinking about everything Miruka had talked about that day.

Non-Euclidean geometries frequently showed up in the math books I'd read, so I'd thought I knew something about them. I'd heard of Bolyai and Lobachevsky. I'd seen many diagrams of spherical geometries and other odd forms. I knew the parallel postulate and what it meant. But the idea of "setting aside" the Pythagorean theorem had never occurred to me. I had seen Escher's works, but I'd had no idea they were related to hyperbolic geometry and the Poincaré disk model. I hadn't realized that infinitely many geometries, infinitely many spaces were available for study through the use of metrics.

Mirka's talk had knocked me off kilter. This wasn't a Tetra-esque "I-know-nothing" game. I really knew nothing!

My ignorance was irritating. I felt as if the entire world was filled with things I didn't know, and that I was unprepared to faced it. I felt keenly aware of my inadequacies, so debilitating that I was only barely able to keep up with my exam studies.

My trembling hand somewhat mechanically reached out and brought my cup to my mouth. I felt the warm tea trickle down my throat.

No, no... no! I can't think like this! Learning is what helps me overcome my weakness! The fact that I'm unprepared is exactly why I'm preparing now!

I recalled that I'd once told Yuri that the math wasn't going anywhere. That there was no need to worry, no need to rush.

With my mindset refreshed, I turned back to my practice problems. I still had tomorrow. I still had my future.

And the math wasn't going anywhere...

> The geometrical axioms are therefore neither synthetic à priori intuitions nor experimental facts. They are conventions.
>
> HENRI POINCARÉ,
> TRANS. BY W.J. GREENSTREET
> *Science and Hypothesis*

CHAPTER 5

Leaping into Manifolds

> I have in the first place, therefore, set myself the task of constructing the notion of a multiply extended magnitude out of general notions of magnitude. It will follow from this that a multiply extended magnitude is capable of different measure-relations, and consequently that space is only a particular case of a triply extended magnitude.
>
> BERNHARD RIEMANN,
> TRANS. BY W.K. CLIFFORD
> *On the hypotheses which lie at the bases of geometry* [25]

5.1 LEAPING FROM THE ORDINARY

5.1.1 *My Turn to be Tested*

First year, second year, third year...

My high school had a strong focus on college prep, so our post-graduation advancement to university had been a more-or-less constant theme throughout the almost three years I'd been there. The pamphlets my school handed out to parents made sure to prominently show data on how many graduating students went on to college. Acceptance rates at selective medical schools. Profiles of students who went on to prestigious national universities...

First year, second year, third year...

For two years I'd watched older students face their own exams, then graduate and leave. Cold temperatures and chilly winds informed me that now it was my turn. There was a gargantuan difference between how the whole exam process appeared from the outside and how it felt from the inside. I'd felt no sense of responsibility as an observer, and I had plenty of freedom to make optimistic predictions regarding what my own experience would be like.

In the midst of the vortex that now surrounded me, however, that was no longer the case. It was like I was wearing blinders, allowing me to see only a sliver of the world around me. I couldn't be sure where I was, or where I was headed. All I could do was plod on and on, ever forward toward the fate that awaited me. *Surprise!* My seniors had all moved on, had all managed to work their way through the storm. But only now that I faced it myself did I realize the enormity of their task.

Honestly, college entrance exams scared me. I found the idea of being tested terrifying.

Even worse, I was disgusted by the fact that it had taken so long to realize such a simple thing about myself.

5.1.2 Defeating Dragons

"Are you even *listening* to me?" Yuri howled, penetrating through my dark thoughts.

It was Saturday, so of course she was hanging out in my room.

"How nice to be you," I sighed. "I'm busy today, Yuri. I have to go over these problem explanations."

"You've already done those problems! What's the point?"

"Just getting an answer isn't enough," I said. "I have to make sure I understand them. To look for better ways to approach similar problems. These explanations help me find my weak points, so I know where I need more work. There's not much point in doing practice problems if you don't take that extra step. You need feedback to improve."

"Sheesh, so serious today. How's your studying going?"

"Not bad, considering this annoying cousin that keeps distracting me."

"Yeah, yeah. Except that all you're doing is working through these problems over and over. Aren't you done yet?"

"I guess I would be, if only I could count on these exact problems being on my exams. Plus I'm also studying for finals in the fall and my practice entrance exams just before Christmas. I really want to do well on my practice exams especially, so I don't have much time." I paused, wondering if this was too much information to dump on a junior high school student. "Don't you have your own mock tests, for your high school entrance exams?"

"I guess? Shouldn't be a big deal though. Junior high is minor leagues." She undid her ponytail and started braiding her hair. "But you? You look like you're gearing up to go battle with dragons."

5.1.3 Yuri's Question

"Dragons, right..." I said. "Anyway, what were you asking me?"

"What four-dimensional dice are," she replied.

"Four-dimensional dice?"

She showed me a page in the book she'd pulled off my shelf, one of my favorite math books from when I was around her age.

"Here. They're talking about how you couldn't bring four-dimensional dice into a three-dimensional world. But I want to know what four-dimensional dice are in the first place."

"Four-dimensional dice are just dice with four dimensions."

"Ha ha, very funny. Now answer my question."

"I did, just think about it. We live in a three-dimensional world, and you know what dice look like here. So what would they look like if you lived in a four-dimensional world instead?"

"And how exactly am I supposed to imagine a four-dimensional world?"

"If I recall correctly, that book tells you exactly how. You just start with three dimensions, then add another. Haven't we talked about this before?"[1]

"Let's pretend we haven't."

"To be fair, no matter how much you talk about objects in a four-dimensional world, you can't really visualize them. So what you

[1] See *Math Girls 5: Galois Theory.*

have to do is use what you know about a four-dimensional world, and think of a four-dimensional die by analogy."

"So what *do* we know about a four-dimensional world?"

"Well, that we can't start from there, for one. We have to start from lower dimensions, and think from there. Here's how we visually understand the first three dimensions from what we learn in school."

I grabbed a piece of paper and made a list.

· 1 dimension — The world of lines

· 2 dimensions — The world of surfaces

· 3 dimensions — The world of solids

"So there you go, three worlds that we pretty much understand. When I was your age, I started here and sort of worked my way up, figuring I could understand the fourth dimension by analogy."

"You were studying stuff like this in junior high? You're a beast."

"Are you sure I haven't told you about all this?"

"Sure enough. So start telling."

And thus began our quest in search of four-dimensional dice.

5.1.4 From Lower Dimensions

"Okay, I'll go through this step-by-step, in the order I thought through them when I was in junior high," I said.

"Sounds perfect," Yuri agreed.

"Okay, so I started with the cubes you find in a three-dimensional world, like the normal dice we're used to. I asked myself, just what kind of shape is a cube? I knew what a cube looks like, of course, but I wanted to delve a little deeper into what it means to be a cube in three dimensions. I figured once I had a good grasp on that, I should be able to transfer my understanding to four dimensions. So after some heavy-for-me-at-the-time thought, I came to the conclusion that *a cube is a collection of joined squares.* That's pretty easy to see if you just imagine a normal die, which has six sides. Each of those faces is a square, and each is attached to four neighbors. So we can think of a cube as a form created from six attached squares."

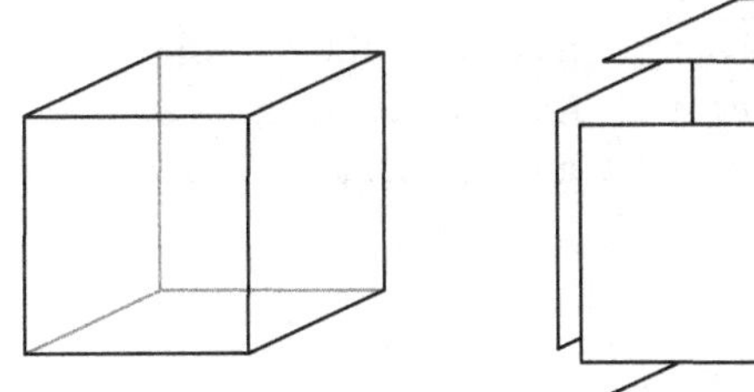

A cube as attached squares.

"Next I made what I considered a really interesting discovery about squares, namely that *a square is a two-dimensional cube*! Which sounds kind of strange, I know, but I decided to run with it and see where it took me. This discovery really excited me at the time, because I realized that instead of describing a cube as attached squares, I could think of three-dimensional cubes as attached two-dimensional cubes. I wanted to understand four-dimensional cubes, remember, but I had no idea what they would look like. If I could create three-dimensional cubes by attaching two-dimensional cubes, though, I'd have taken a big step in that direction, because I could just go up one dimension and think of four-dimensional cubes as attached three-dimensional cubes.

"So if I wanted to create a four-dimensional cube, all I had to do was attach a bunch of three-dimensional cubes, in other words normal dice shapes. I was sure this was the right approach, because it seemed correct, and natural, and elegant. There was a certain consistency to it."

"Well done, sir, well done," Yuri said.

"I got really into this line of thought, but it kind of scared me."

"Scared you? Why?"

"Because maybe I was wrong," I said. "It was just some crazy idea I had, so I wanted some way to verify it as being truly true."

"Ah."

"So, I decided to try going down a dimension instead. I thought, if a three-dimensional cube is the normal shape we're used to, and a two-dimensional cube is a square, what would a one-dimensional cube be? If I attached some one-dimensional cubes, could I create a two-dimensional cube? Of course, the answer to that wasn't hard. A one-dimensional cube would be an edge of a two-dimensional cube, in other words a line segment. And sure enough, by assembling some one-dimensional cubes, some line segments, I could create a two-dimensional cube, a square!"

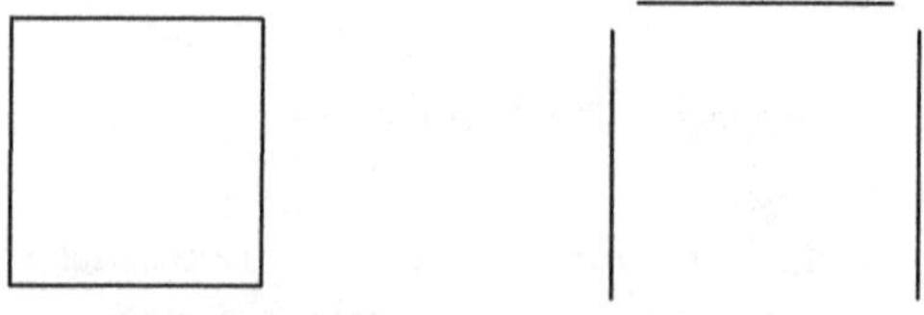

A square as attached line segments.

"So again, excited. I'd confirmed that when going down one dimension, I could consider two-dimensional cubes as being attached one-dimensional cubes. But then, I realized I had a big problem."

"That being?"

"The *interiors* of all these shapes, and what they contained. I realized I could make two kinds of dice. One would be like something I made out of clay, where the interior would be filled with stuff. But I could also create an origami-like cube, which would have a surface but nothing inside."

"Yeah, sure, but . . . A cube's a cube, right? Is what's inside really important?"

"It is! I'd started from the idea that a cube is a collection of squares, and that led me to the empty kind of cube, but one made from squares that are filled with stuff, since they're made from, like, pieces of wood or whatever. See how that creates a gap?"

"Ah, I see. You're attaching *filled* two-dimensional cubes to create *empty* three-dimensional cubes. And that was the problem, right?"

"Right. I wanted everything to build up by analogy, so I couldn't freely jump back and forth between filled and non-filled objects."

"Quite the quandary. So how did you resolve it?"

"By differentiating between filled objects and those that were empty, like this."

- Call filled objects *die solids*, and

- call objects with surfaces but nothing inside *die surfaces.*

"With this, I needed to modify my discovery from before."

- A one-dimensional die surface is four connected one-dimensional die solids (we create the periphery of a square from four filled line segments).
- A two-dimensional die surface is six connected two-dimensional die solids (we create the surface of a cube from six filled squares).

"Very interesting!" Yuri said.

"So finally, I figured, I had the proof I needed. I was ready to dive into the four-dimensional world, like this."

- A three-dimensional die surface is some number of connected three-dimensional die solids!

"Way to go!" Yuri yelled. "That's— Wait, kinda weird? I thought you wanted to make four-dimensional dice, not three-dimensional ones?"

"We have to think carefully here. With the way I'm naming things, a three-dimensional die surface is fine. Remember, an origami-like die is a *two*-dimensional die surface, right? A die having only surfaces can exist in a three-dimensional world, but it's still what we're calling a two-dimensional die surface."

"Ah, okay. Sure."

"So when we take that up one dimension, we definitely have a *three*-dimensional die surface. What we want to consider here is a three-dimensional die surface sitting in a four-dimensional world."

"Oh, all right!"

"Next, I started thinking about how we can attach three-dimensional die solids."

"Hold up, I think I know!" Yuri had a gleam in her eyes. "I know how to connect three-dimensional die solids to create a three-dimensional die surface!"

"Do you now?"

Yuri pulled the paper over in front of her, and started speaking and writing carefully to ensure she was getting everything right.

"If you want to know this..."

- How can I attach three-dimensional die solids to create a three-dimensional die surface?

"...you just do like you did before and ask yourself this."

- How can I attach two-dimensional die solids to create a two-dimensional die surface?

"Well done, Yuri! That's right, you just think in a lower dimension."

Yuri blushed, but patted herself on the back. "More praise, keep it coming."

"Nah, don't want it going to your head."

"Grrr. Anyway, let's see... A two-dimensional die solid is a filled square, right? And to create a two-dimensional die surface, you attach a bunch of squares so that each of their sides perfectly lines up with another square."

"Go on."

"But that's it, right? You create a two-dimensional die surface by attaching adjacent squares."

"That's right. We gather six two-dimensional die solids, attach their sides, and we have a two-dimensional die surface. Here's what that would look like, with one of the two-dimensional die solids colored."

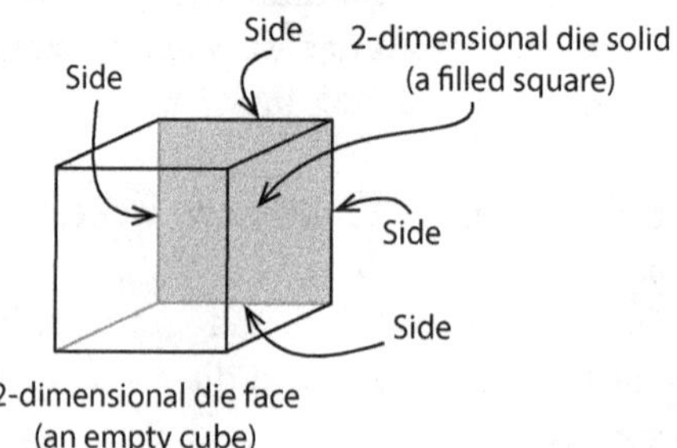

Connecting sides of two-dimensional die solids.

"Okay," Yuri said, "so we just do the same thing using three-dimensional die solids, right? A three-dimensional die solid has six faces, so to make a three-dimensional die surface we just perfectly attach adjacent three-dimensional die solids to those six faces... But we can't do that!"

"Why not?"

"Because a three-dimensional die solid is a cube filled with stuff, right? When you get a bunch and stick them together, you don't want any faces that aren't attached, but that's impossible! You'd have to, like, bend stuff all around."

"That's exactly right, just as you say."

"So that's a no-go, right? If everything gets all bendy twisty, it isn't a cube anymore!"

"Not in three dimensions it isn't."

"Huh?"

5.1.5 Bendy Twisty

"You're right that we're trying to attach each the sides of a three-dimensional die solid," I said, "in other words a die that's filled with... stuff. We want to attach the faces of this filled cube to each other, to create a three-dimensional die surface situated in a four-dimensional space. That's hard to picture in just three dimensions. All we can do is warp the shape."

Yuri cocked her head. "Yeah? But then we don't know what shape we're dealing with."

"That's exactly what I first thought when I was working through this. But then I realized, we're *always* looking at distorted dice."

"How so?"

"Well, what we're trying to do is view, in three dimensions, a three-dimensional die surface situated in four dimensions. So once again, let's take things down a dimension. In other words, think about viewing in two dimensions a two-dimensional die surface placed in three dimensions. How would we do that?"

"If you want to see something in two dimensions, I guess you could just draw it on paper?"

"Good idea. If we draw a two-dimensional die surface in two dimensions, we get something like this."

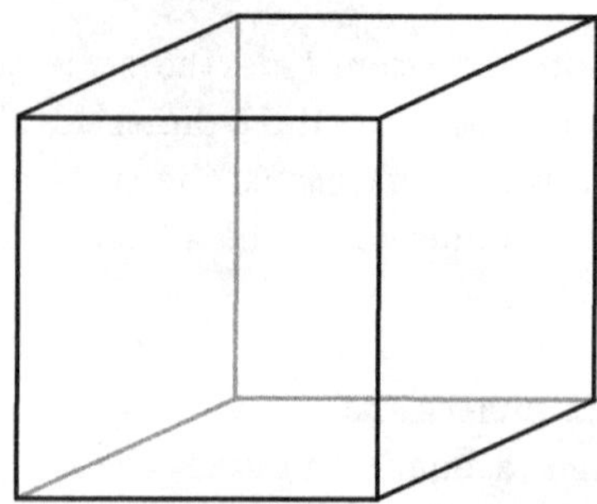

A two-dimensional die surface viewed in two dimensions.

"This isn't warped," Yuri said. "They're all squares."

"You sure about that? I think you might be changing things back into squares in your mind. There's supposed to be six squares here, but the only ones I've actually drawn as squares are the front and rear sides. The other sides are slanted into parallelograms. That's what I mean by the shape being warped. It's like this."

- When viewing in two dimensions a two-dimensional die surface situated in a three-dimensional space, the two-dimensional die solids creating the two-dimensional die surface appear distorted.

"So, it makes sense that ..."

- When viewing in three dimensions a three-dimensional die surface situated in a four-dimensional space, the three-dimensional die solids creating the three-dimensional die surface appear distorted.

"Oh ho!" Yuri said. "Okay, I'm good up to this point. But do it for real, even if it is distorted or whatever. Show me what a three-dimensional die surface made from three-dimensional die solids would look like."

"It would look something like this."

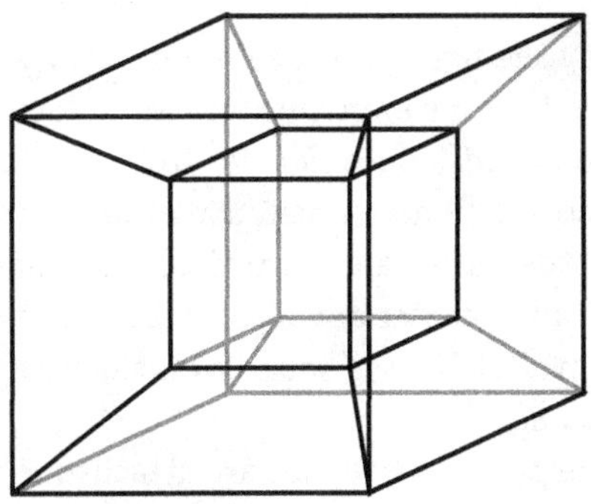

A three-dimensional die surface viewed in three dimensions.

"What's this?" Yuri asked.

"Yeah, it takes some imagination, I know. Actually, this is like a three-dimensional model, but drawn on paper. Strictly speaking, it's a four-dimensional object dropped down to three dimensions and drawn two-dimensionally."

"Wow, that's deep."

"There's eight dice here. No matter which one you look at, its faces are attached to faces of adjacent dice faces. The easiest one to see is the small die in the middle. See how all six faces there are attached to surrounding, distorted dice faces? Those distorted dice look kind of like pyramids with their tops chopped off. There's six of those pruned pyramids in this figure, each facing a different direction. For example, here's one."

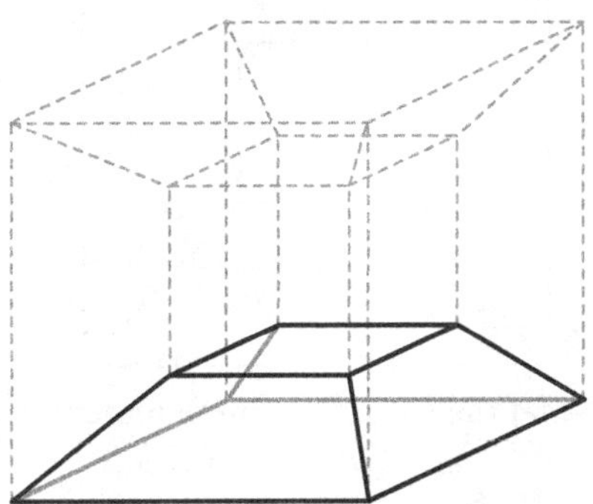

One shortened pyramid.

Yuri looked at the figure I'd drawn without speaking, clearly in deep-think mode. The autumn sun created golden sparkles on her chestnut braid. I waited quietly for her to return to our plane.

Finally, she spoke. "There's still something weird about this. I understand how the cubes are warped and all, and I see how the small cube faces in the center are all connected to the headless pyramids around them. But the big squares on the base of each pyramid aren't attached to anything!"

"They are. The pyramid bottoms are attached to the biggest cube on the outside."

"What biggest cube on the outside? All the other cubes are on the inside here. So there's only seven in all, right? One here in the middle, and the six stubby pyramids around it."

"You're following exactly in my footsteps. That's exactly what I thought, first time I saw this. But no, there are eight dice in this figure. The small one in the middle, the six truncated pyramids, and the biggest cube, which surrounds them all."

"Not seeing that."

"It's easier to see if you compare this with another way to draw a two-dimensional die surface on paper."

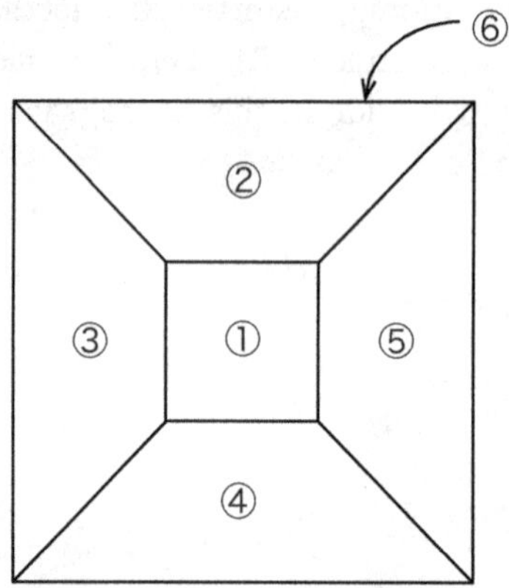

A two-dimensional die surface in three dimensions, compressed into two dimensions.

"Imagine one square on a die is stretched out and smooshed flat," I said, "so we can compress this two-dimensional die surface in a three-dimensional space down into the two dimensions of the page. That's the big square on the outside labeled 6 in the diagram."

"Ah, okay. I see that."

"Now just think of the diagram with the three-dimensional die surface in the same way. In other words, we've taken a three-dimensional die surface in four dimensions and smooshed it down to three dimensions, resulting in one cube looking larger and surrounding the rest."

"Okay," Yuri begrudgingly said. "But I still say there's something weird about these inside–outside shapes..."

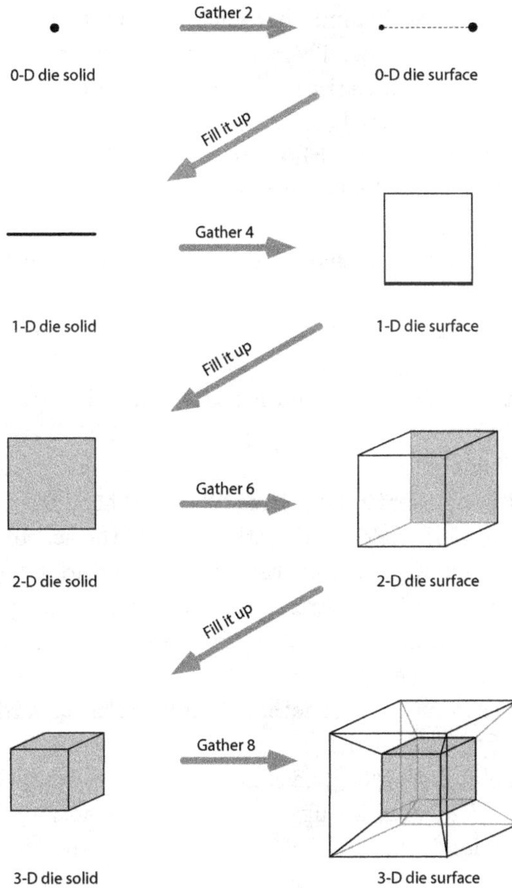

5.2 Leaping into the Unordinary

5.2.1 Beneath a Cherry Tree

On Monday morning, we had a school-wide assembly. Every week we would all gather in the auditorium, where the principal delivered unto us his wisdom. We were now deep into autumn—the perfect season for focusing on our studies and the end stretch for those of us who would be graduating—so it was grating to have to sit there listening to his blather.

I pretended to adjust my glasses to hide a yawn.

My first-period class was just a study hall, so after the assembly ended I slipped away behind the school. I found myself strolling down a row of cherry trees. There was no one else around.

I found myself by a particularly large cherry tree. *The* cherry tree. I looked up into its bare branches. Had this been spring, it would have filled my view with pastel pinks, blotting out all other existence. But now, in the fall, it was just another tree.

"Remember that day?"

I spun around at the sudden voice, and saw Miruka standing there.

"Of course," I replied. We were in the exact spot where we'd first met, back in the spring of my first year.

Miruka stood next to me and looked up into the tree. A citrus scent tickled my nose.

"I do too," she said.

We stood there silently for a time. Shouts echoed from far away, probably the athletic field on the other side of the school building. Some PE class, forcing its students out into the cold. They seemed a far fainter presence than Miruka, myself, and the cherry tree.

The silence became unbearable. "So I guess you've found your shortest path out of here," I said.

Miruka turned her gaze (another shortest path) toward me.

"I want coffee," she said.

I thought of my book bag, back in my classroom where I'd left it, but shrugged. Wherever Miruka was willing to lead me, I followed without hesitation.

5.2.2 Inside Out

We entered Beans, a coffee shop right next to the train station, and took seats across from each other.

"Studies coming along?" Miruka asked, causing me to grate my teeth.

"Well enough," I replied. "But I don't want to talk about that. I get enough from my mother."

"But much more sweetly, I'd imagine." Miruka lifted her coffee up to her face, slightly steaming her glasses. "How's she doing, by the way?"

"Looks like she's going to be fine."

"And Yuri? I haven't seen her in a while."

"Same as always. Reading books, doing math." I gave her a brief rundown of our recent discussion of four-dimensional dice. "...So in the end we were able to create a three-dimensional die surface by combining eight three-dimensional die solids. I don't think Yuri's quite satisfied yet, though. Something about faces overlaying each other when we lower the dimensionality is still bugging her."

"Hmph. Try adding a point at infinity and turning everything inside out."

"Inside out?"

Miruka made writing gestures.

I knew I should have brought my book bag, I thought, approaching the barista to borrow a pen and a notepad.

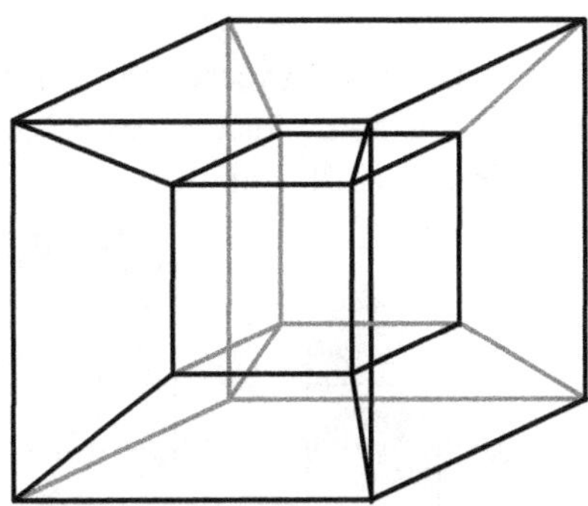

A three-dimensional die surface viewed in three dimensions.

"In this diagram, you can think of the big cube as being turned inside out," Miruka said.

I cocked my head. "How so?"

"You're considering everything on the outside in this figure as being 'outside' the cube. But try thinking of it as being inside the cube instead. Then you've captured the entire cosmos within your three-dimensional die solid."

"I have no idea what you're talking about."

"Okay, consider space as being infinite. Floating within it is a cube, made of glass. At this point, everything in the surrounding space is contained within this eighth cube. Then the inside-out cube containing all of space has six square faces, each attached to the bottom of the six pyramids."

"Wah?" I sputtered, making an odd moan. *This is . . . This is . . .*

"Seeing it now?"

"I see it," I managed. "The cube . . . Flipped inside-out . . ."

"There you go. That's one way we can represent a three-dimensional die surface being crammed into a three-dimensional space. Topologically speaking we need to add a point at infinity, though."

"This is . . . heady stuff."

"Oh, we can also do the same thing when lowering the dimensions. Want to try forcing a two-dimensional die surface into a two-dimensional space?"

"I know that one. One of the squares gets bigger when you cram it in, right?"

"When you do it that way the square overlaps, so you're considering its entire periphery as being 'inside.' Everything in the region labeled 6 here is 'inside' the sixth square."

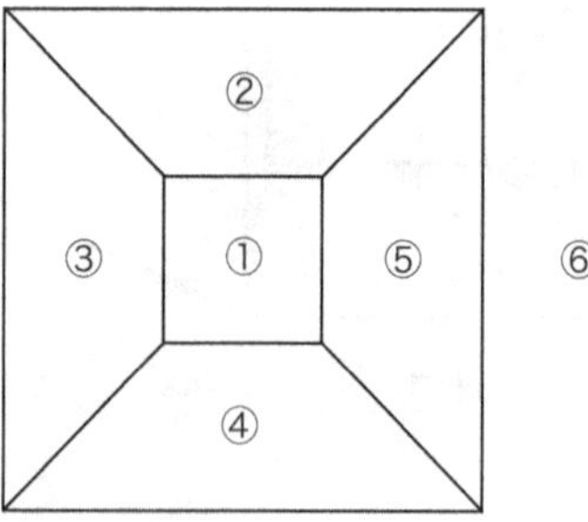

"Even in two dimensions, heady stuff."

5.2.3 Nets of Polyhedra

"Smooshing things down and turning them inside-out is fun and all," Miruka said, "but there's a way to do this without all the distortion, you know."

I looked up from Miruka's diagram. "Oh yeah?"

"Just create a net for your three-dimensional die surface." She began sketching. "But let's warm up with a net for the two-dimensional version. We start by detaching some of the edges, then spread them out. That gives a net in the plane with no distorted squares, the downside being that the edges we sliced open now appear in two places, so we should draw some arrows to indicate edges that are actually connected."

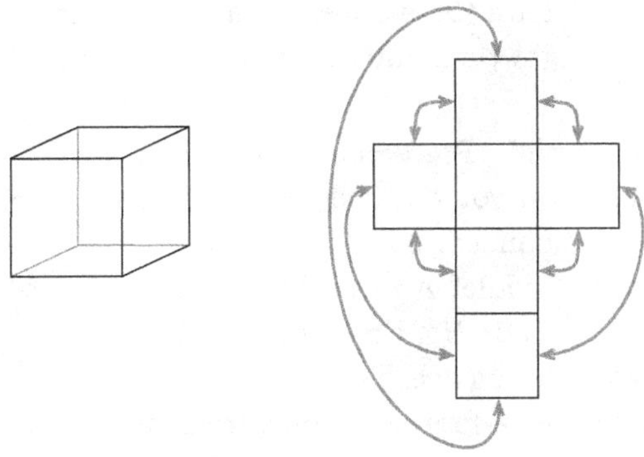

Net for a two-dimensional die surface
(six two-dimensional die solids).

"Interesting. So now we want to bump this up to a three-dimensional die surface, right? The net for a two-dimensional die surface was a collection of squares, so I guess the net for a three-dimensional die surface would be a collection of cubes."

Miruka nodded. "That's right. Same process, just with everything bumped up a dimension. To create a net for a three-dimensional die surface we separate some connected faces, and spread those out. So now we have an expansion into cubes with no distortion, but with the same downside—connected faces are now shown in two places. So again, arrows."

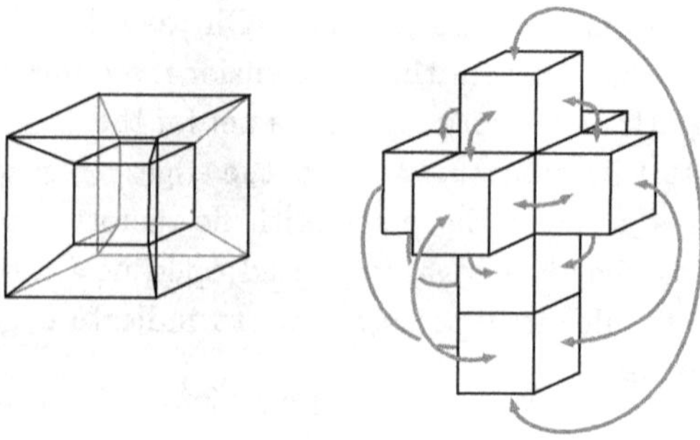

Net for a three-dimensional die surface
(eight three-dimensional die solids).

"Cool stuff," I said. "Pretty sure I get this."

"Do you 'get' how you can make another leap to imagine this as something that's bounded, but without end?"

"What's that, a riddle? A three-dimensional die surface is clearly finite, but without end? No, I'm not seeing that."

"Imagine you lived on one. Then you could pick any direction and keep heading that way, forever, leaping from cube to cube. Imagine going from one face in a cube to its opposing face, then to the next one and the next. What happens if you keep doing that?"

I sat there for a moment, staring at the net Miruka had drawn.

"Ah, okay. After you've gone through four cubes, you end up back where you started."

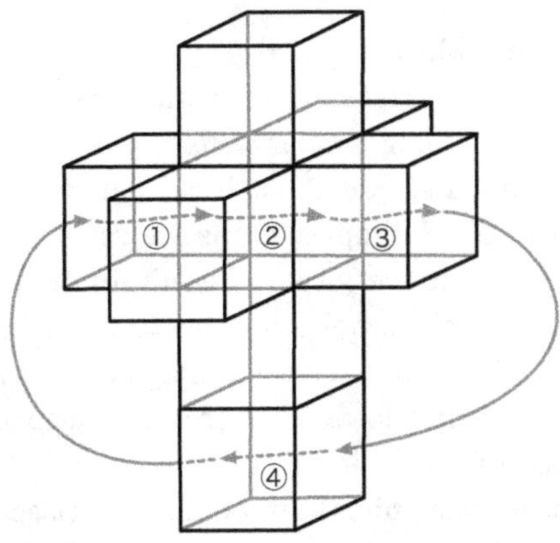

An infinite walk through a three-dimensional die surface.

"That's right," Miruka said. "You'd feel like you were always heading in a straight line, but you would just be going around and around within a finite space, in this case within four cubes. Residents of a three-dimensional die surface would be trapped there, unable to find the 'outside.' No matter what direction they moved in, they would just find themselves moving through attached faces into another adjacent cube."

"Just like residents of a two-dimensional die surface would be trapped within their squares, I guess."

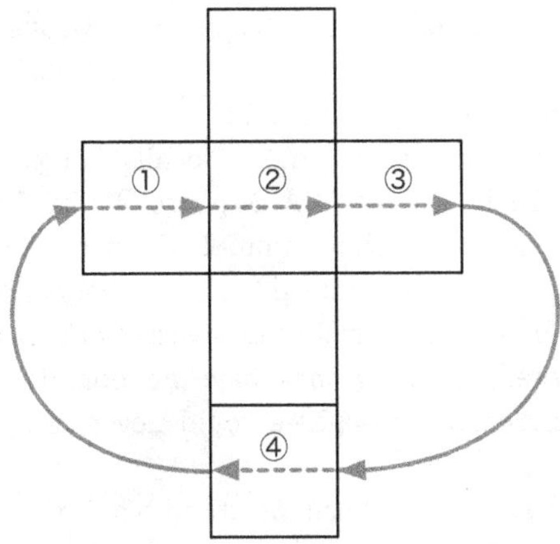

An infinite walk through a two-dimensional die surface.

"Exactly like that." Miruka smiled.

"This is fun, imagining traveling through these weird spaces."

"Yeah? We did something similar before, tracing along the surfaces of Möbius strips and Klein bottles. That was movement in two dimensions too. But I suppose while we can talk about 'tracing' two dimensional surfaces, that doesn't quite carry over to three-dimensional ones, since we're so used to being inside a three-dimensional space. We would have to imagine ourselves as four-dimensional beings, looking down on a three-dimensional space from the outside, tracing a finger along it."

"Except that truly imagining that would be just as difficult as a two-dimensional being imagining viewing its two-dimensional space from the outside."

"Saying a world is without boundary means you can walk in any direction and never come to an edge," Miruka said, becoming excited. "But that comes in two flavors—a world that's unbounded without boundary, or one that's bounded without boundary. Euclidean planes and spaces are the first type, while two- and three-dimensional spheres are the second."

"You sure about that?" I said. "I can see how a two-dimensional being would be trapped on its two-dimensional sphere, but a three-dimensional being could leave its sphere, right?"

"That would be quite a trick," Miruka deadpanned. "But I think you're misunderstanding something about spheres with a three-dimensional surface, also called a 3-sphere. A 3-sphere is a type of 3-manifold."

"Okay, I'll bite. What's a manifold?"

"An n-manifold is a space that is locally homeomorphic to an n-dimensional Euclidean space. A 2-sphere is a kind of 2-manifold, and a 3-sphere is a kind of 3-manifold. From every point on a 3-manifold, if you survey your neighborhood everything will appear to be a three-dimensional Euclidean space. Both 2-spheres and 3-spheres are closed, meaning they have no boundary, so a three-dimensional resident of a 3-sphere could never escape to 'outside' of it."

"Uh... So have I really misunderstood what a 3-sphere is?"

"Hmph. Ever heard of the Poincaré conjecture?"

5.2.4 *The Poincaré Conjecture*

"The Poincaré conjecture?" I said. "Sure, I've heard of it. I saw something about it ... uh ... on a television show."

"The Poincaré conjecture involves a 3-sphere called S^3," Miruka said. "A common confusion is that we're talking about the 2-sphere S^2 instead, since when you hear something about three-dimensional spheres it's easy to imagine an object something like the surface of a ball, even one filled with something."

"That's, yeah ... exactly what I was imagining."

"Despite the fact that you were calling your three-dimensional die surface something similar?"

"Ah, right," I said, now seeing where I was getting confused. "Maybe because in that TV show I watched they showed something like a ball wrapped in string. That visual stuck with me somehow."

"That was their representation in a lower dimension. The ball's surface is a 2-sphere, which is completely different from a 3-sphere. I mean, a two-dimensional die surface is completely different from a three-dimensional die surface, right?"

"Indeed it is."

"So the 3-sphere that shows up in the Poincaré conjecture is *not* the surface of a ball, and it doesn't have stuff inside of it. Actually, 'space' is the best everyday term that describes the 3-sphere."

"Wow, okay. But still, I can imagine a three-dimensional die surface as distorted cubes, or as a net of polyhedra, but this? I don't think I can wrap my brain around it."

"The 3-sphere is a 3-manifold that's homeomorphic to your three-dimensional die surface. We can think of a 3-sphere in the same way as your three-dimensional die surface."

5.2.5 *The 2-sphere*

"Before we jump into 3-spheres, however," Miruka said, "let's get a bit more comfortable with 2-spheres.

"Imagine a rubber balloon. Even better, a balloon globe of the earth. That's a 2-sphere. Separate that globe at the equator, and spread out and flatten the northern and southern hemispheres. That gives you two disks. If you reattach the disks, you'll do so along the equator, which is a 1-sphere."

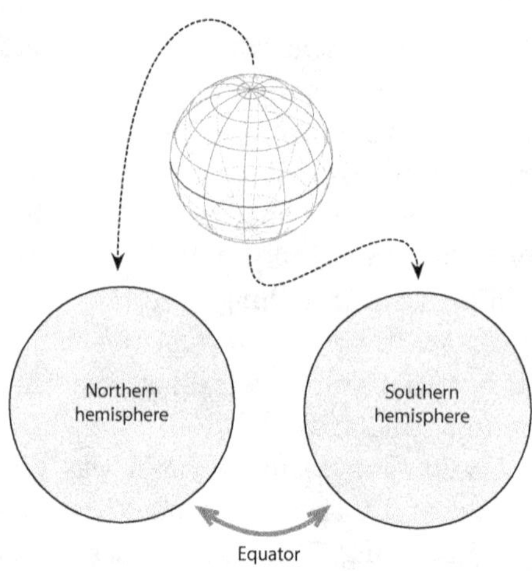

The surface of a globe split into two disks.

"A two-dimensional resident here could move on the globe, from the northern hemisphere and crossing the equator to the southern hemisphere. If that critter kept moving, it would eventually cross the equator again, back into the northern hemisphere. It can keep going on and on, but clearly it's in a bounded space. This is similar to living on earth, which we experience as bounded but without boundary."

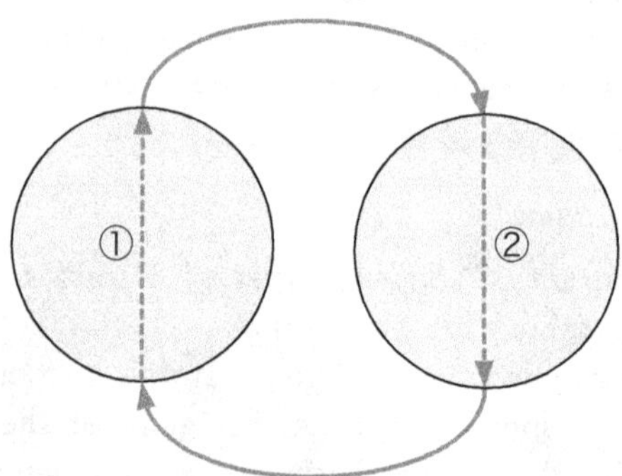

A two-dimensional entity traveling across a 2-sphere.

5.2.6 *The 3-sphere*

"I'm good with the 2-sphere," I said. "Cut along the equator and spread out into two disks, no problem. But a 3-sphere...?"

"Well," Miruka said, "we've just seen how two filled disks joined at their peripheries is homeomorphic to a 2-sphere. So just bump everything up a dimension and do the same thing. In other words, take two balls, whose boundaries are spheres, and attach their surfaces to create something homeomorphic to a 3-sphere."

"Wait, we're attaching the surfaces of two balls?"

"Sure, just takes a little imagination. Before we had a two-dimensional creature taking a trip across the equator, which represents the seam between two disks. Now we have a three-dimensional creature taking a trip across a spherical surface, which represents the connection between two balls."

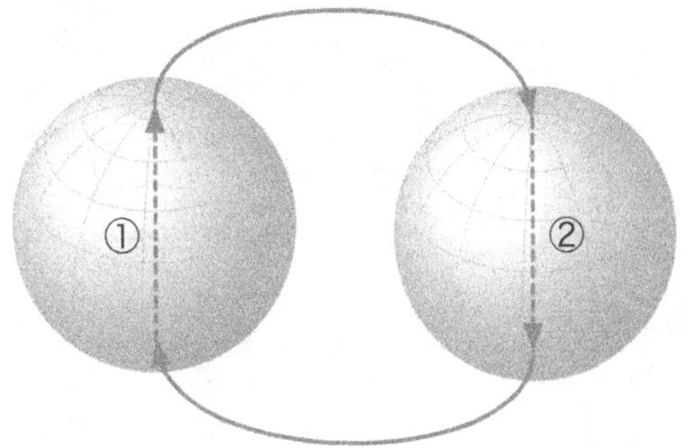

A three-dimensional entity traveling across a 3-sphere.

"Gimme a second to process this," I said. "So we start out moving 'inside' one ball, then we cross its surface to enter 'inside' the other ball?"

Miruka nodded. "Exactly. This diagram only shows one loop, but be sure to remember that no matter where on the surface of the first ball you emerge from, you're simultaneously entering the 'inside' of the other ball."

"Ah, of course. Because we said the surfaces of both are attached. Okay, I think I understand what that means now."

"Then you're well on your way to imagining a 3-sphere as two balls. Generally, we write an n-sphere as S^n, and we can consistently represent S^n as equations, like this."

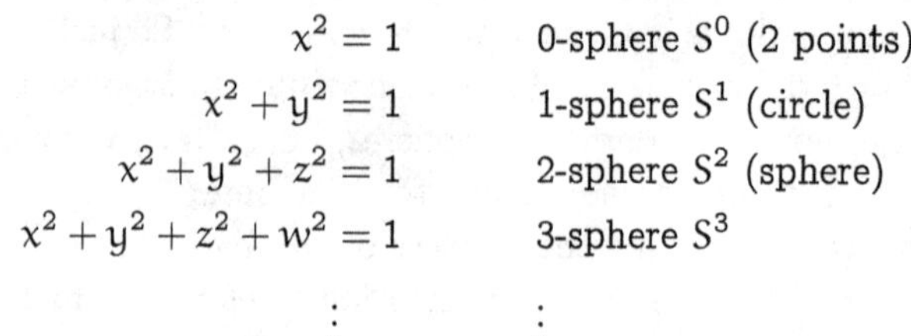

$$
\begin{array}{rl}
x^2 = 1 & \text{0-sphere } S^0 \text{ (2 points)} \\
x^2 + y^2 = 1 & \text{1-sphere } S^1 \text{ (circle)} \\
x^2 + y^2 + z^2 = 1 & \text{2-sphere } S^2 \text{ (sphere)} \\
x^2 + y^2 + z^2 + w^2 = 1 & \text{3-sphere } S^3 \\
\vdots & \vdots
\end{array}
$$

Attach 2-spheres (spheres)

3-ball

Net for a 3-sphere

Attach 1-spheres (circles)

2-ball

Net for a 2-sphere

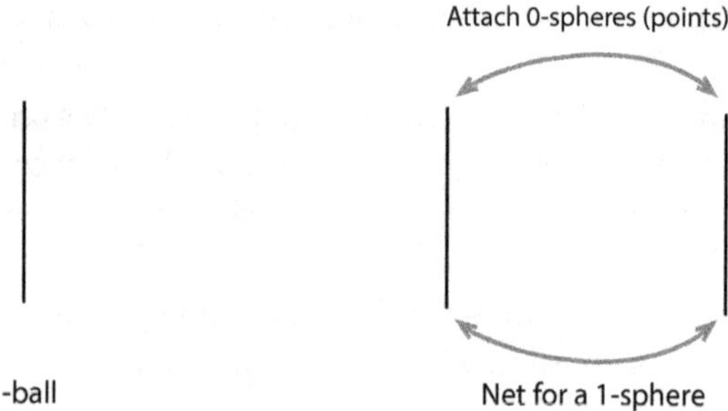

1-ball

Net for a 1-sphere

"Before you were talking about n-manifolds, right?" I said. "You said something about a 3-sphere being a kind of 3-manifold."

"I did," Miruka said, turning her eyes to me and taking a sip of her now cold coffee. "What of it?"

"You said a 3-manifold is locally homeomorphic to Euclidean space, which I think means that, topologically speaking, we couldn't distinguish between looking around our location in a three-dimensional Euclidean space and doing the same in a 3-sphere. Am I right there?"

"Roughly speaking," she replied. "But don't forget the 'locally' part. A 2-sphere is locally homeomorphic to two-dimensional Euclidean space, but its entirety isn't. Same for a 3-sphere and three-dimensional space."

"I guess I'm still stumbling on that, how we can't differentiate between what kind of space we're in locally, but we can globally. I mean, if you're looking at something 'globally,' aren't you looking from the outside? I guess I can manage two-dimensional curves, but I have no idea how to imagine jumping into a space and viewing something in its entirety in three dimensions, much less anything higher than that. But still, you're saying if I fell asleep in a Euclidean space and woke up in a 3-sphere, I wouldn't notice anything was amiss?"

"How poetic. Actually, there is one tool you could use to get a better grasp on your surroundings."

"That being?"

"Groups. But let's save that discussion for later."

And so we did.

5.3 Jumping in, Leaping out

5.3.1 Waking up

"Um, hi?" came a voice, causing me to jerk up.

I saw a girl with short hair and big eyes before me, peering at my face with some concern.

"Tetra?" I said.

I scanned my surroundings. Many desks. Familiar bookshelves along the walls. In the distance, a bronze statue of some philosopher.

Next to it, a wheeled cart stacked with books. Of course. The school library.

"Hello there?" Tetra repeated. "Sorry to bother you, but school's almost closed, so..."

"It's that late already?" I said. I glanced at the window, and sure enough it was getting dark.

I had been working on practice problems after classes, and I'd gotten so involved in my calculations that I'd blotted out the rest of the world. Until Tetra brought me back, at least. Being ejected from my realm of thought back into the real world had knocked me for a loop.

Or had I been sleeping?

"Ms. Mizutani should be appearing soon," Tetra said, clenching and spreading her hands. "So I figured, y'know,—"

"Sure," I said. "Let's head back together."

"Great!"

5.3.2 Eulerians

Walking together toward the station, I realized it had been some time since I'd been alone with Tetra. It had become too cold to eat lunch on the school roof, and Miruka had been present for our recent math talks. The main reason, though, was that I'd been spending a lot more time alone, studying for exams.

As we wound our zig-zagging path through the residential area between our school and the station, Tetra suddenly spoke up.

"I've created the Eulerians!"

"Right, you said something about that the other day. So what exactly are they?"

"A math club!"

"A math club? As in, you're going to get a bunch of people together to do math?"

"That's right! Well, Lisa and I are the only members so far. But it's our new group, a group for Euler fans and aficionados. Lisa's doing a lot of programming to create some cool graphics like Poincaré disks. I'm hoping I'll be able to do stuff like that eventually, but I'm not quite there yet. So for now I'm in charge of writing!"

Tetra was communicating excitedly, with both words and gestures, but I still wasn't entirely sure what she was going on about.

"Writing?" I said.

"Oh, right. We're also going to produce a magazine called *Eulerians*. Like, a group newsletter, or a booklet, or... well, something like that. I'm not sure how many pages it'll have, but anyway we're going to write articles about what we're thinking about and what we've done, and print them up!"

So that's *what she's been working on...*

"Sounds cool," I said.

"I just want to put what I've learned into a physical form," Tetra continued, obviously excited about her new project. "Remember when I gave that presentation on quicksort[2] last year, at the Narabikura Library? There were so many people in the auditorium that day, so many other students. I really fumbled—I was so nervous—but still, that was a really important experience. So many people seemed to enjoy my talk, and I learned so much preparing for it. It was just so... satisfying. It made me satisfied with who I am."

"Wow," I said, searching for better words, but failing to find them.

"But still... the only people who got to hear my presentation were those there in the auditorium. I did make handouts to cover stuff that wasn't in my talk, but I didn't have time to really go into much detail in those, so they aren't really useful for anyone else. So in a sense, it's like everything I did that day is just... scattered to the winds."

Tetra spread her hands as if strewing her ideas into the night sky above us. Or perhaps she was making an appeal to the entire world.

We reached an intersection and stopped to wait for the light to change. I remained silent, but Tetra resumed her exaggerated gestures to emphasize how she was feeling.

"I want to give form to what I've learned, to what I'm thinking about. But there are limits to what I can do alone. That's what prompted me to work with Lisa to create the Eulerians. She was so helpful with my presentation, after all."

The light changed and we started walking again. Tetra continued talking excitedly, but I was still at a loss for words. I knew I should

[2]See *Math Girls 4: Randomized Algorithms.*

tell her how wonderful her idea was, that I should give her some encouragement, but somehow nothing came out.

"Oh, and I came up with a new hand sign, too!" she said as we passed through a deserted park. "You just— Hey, are you okay?"

Apparently she had noticed my muteness. She stopped and looked up at my face.

"I..." I searched for something, anything to say. *I can't bring myself to be happy for you.* But what I managed was, "I just don't have the... capacity to deal with this right now." *Not what I wanted to say...*

"Oh, no. No, I'm sorry. I wasn't trying to recruit you or anything. I know you're busy with your studies and all. I just—"

"I'm weak," I said.

"You're— What?"

"I'm scared."

"I... I don't—"

"I want satisfaction too. I'm limited and I'm weak and I'm scared, but... I want satisfaction too. But all I've been thinking of is, like, getting into a good school, trivial things like that." I slumped onto a park bench beneath a light and continued talking, half to myself. "I think your project, the Eulerians and your magazine and all, sounds really great. Best of luck on that. But it's kind of painful at the same time, because my own goals seem so pitiful in comparison."

Tetra sat next to me. I felt her hand on my back, and smelled her sweet scent.

"C'mon now," she said. "That's no way to talk. I've learned *so much* since I met you and Miruka. You guys have also shown me how interesting it is to learn new things—how fun, and how beautiful, how moving. You're the reason I now want to pass these feelings on to others. It's only because you've shown me so much that I want to learn so much more. So please, don't talk like that."

I could hear the tears in Tetra's words, and a slight trembling in the hand on my back. I looked up at the night sky and saw stars.

Stars appeared to me as forever rotating about a fixed point, but in truth it was I who was forever spinning, around and around. I was simply thrashing about in this space I'd been thrust into, forever swirling around the same place.

In the extension of space-construction to the infinitely great, we must distinguish between *unboundedness* and *infinite extent*; the former belongs to the extent relations, the latter to the measure-relations.

BERNHARD RIEMANN,
TRANS. BY W.K. CLIFFORD
On the hypotheses which lie at the bases of geometry [25]

CHAPTER **6**

Capturing Unseen Forms

> The proposed goal is to determine the characteristics for the solubility of equations by radicals. We can affirm that in pure analysis there does not exist any material that is more obscure and perhaps more isolated from all the rest.
>
> ÉVARISTE GALOIS
> TRANS. BY P.M. NEUMANN

6.1 CAPTURING FORM

6.1.1 The Form of Silence

Before a race, Formula One drivers perform a sort of self-tuneup. I always did the same before taking a big test.

I used the restroom. I did some light stretches. I arranged my pencils and eraser and a small digital clock on my desk. With everything thus prepared, I felt sure I could fully focus. By taking practice tests like the one I was taking that day, I'd become accustomed to this pre-test regimen. There was one thing I never got used to, though—the silence. The silence that fills the room before the test begins.

Proctors would walk around the room, handing out answer sheets. We test-takers would follow their progress, in our consciousness if not with our eyes. The only sound would be a nervous cough here and there, but our minds would be racing. It was that noisy silence I couldn't get used to. Once the test got started it was no big deal, because my brain would be fully occupied. But in that time before I could engage myself with problems, it would just be spinning its wheels. Given nothing to work on, it would fill itself with useless thoughts. Things like how I'd acted with Tetra the other night in the park. How I'd made such a fool of myself in front of her.

At school the next day she'd smiled at me, acting as if nothing had happened. She'd been almost angelic. An exaggeration, perhaps, but she'd remained sincere and energetic, and she listened intently to everything I said. A little bit flighty, as usual, but also innocent as usual.

A bell rang in the classroom, and everyone opened their test booklets.

6.1.2 The Form of Problems

Problem 6-1 (a recursion formula)

Let $\theta = \frac{\pi}{3}$, and define a function

$$f(x, y) = (x\cos\theta - y\sin\theta, x\sin\theta + y\cos\theta)$$

mapping the real number pair (x, y) to the real number pair $(x\cos\theta - y\sin\theta, x\sin\theta + y\cos\theta)$. Further, let sequences $\langle\, a_n \,\rangle$ and $\langle\, b_n \,\rangle$ have elements recursively defined as

$$\begin{cases} (a_0, b_0) & = (1, 0) \\ (a_{n+1}, b_{n+1}) & = f(a_n, b_n) \qquad (n = 0, 1, 2, 3, \ldots). \end{cases}$$

Find

$$(a_{1000}, b_{1000}).$$

There's nothing like the combination of a complex problem and a time limit to put you off balance. It's only natural. But keeping calm and focusing on the form of the problem will reveal a solution. Same thing here.

The key to this problem was to realize the form that this bit described:

$$f(x, y) = (x\cos\theta - y\sin\theta, x\sin\theta + y\cos\theta).$$

I could think of this mapping f as moving a point (x, y) in the coordinate plane to another point $(x\cos\theta - y\sin\theta, x\sin\theta + y\cos\theta)$. In doing so, it would rotate the point θ degrees around the origin. I'd played with this many times before. So many times, in fact, that I felt more comfortable writing it as a matrix, like this.

$$\begin{pmatrix} \cos\theta & -\sin\theta \\ \sin\theta & \cos\theta \end{pmatrix}$$

The product of this matrix and a column vector would be

$$\begin{pmatrix} \cos\theta & -\sin\theta \\ \sin\theta & \cos\theta \end{pmatrix} \begin{pmatrix} x \\ y \end{pmatrix} = \begin{pmatrix} x\cos\theta - y\sin\theta \\ x\sin\theta + y\cos\theta \end{pmatrix},$$

so we were taking some point (x, y) and moving it like this:

$$\begin{pmatrix} x \\ y \end{pmatrix} \stackrel{f}{\longmapsto} \begin{pmatrix} x\cos\theta - y\sin\theta \\ x\sin\theta + y\cos\theta \end{pmatrix}.$$

Points rotating around the origin, like stars around a pole star. Having made it this far, the rest should be easy.

So the problem is asking, what happens when you apply this mapping f one thousand times?

$$\underbrace{\begin{pmatrix} 1 \\ 0 \end{pmatrix} \stackrel{f}{\longmapsto} \begin{pmatrix} a_1 \\ b_1 \end{pmatrix} \stackrel{f}{\longmapsto} \cdots \stackrel{f}{\longmapsto} \begin{pmatrix} a_{999} \\ b_{999} \end{pmatrix} \stackrel{f}{\longmapsto}}_{\text{apply f 1000 times}} \begin{pmatrix} a_{1000} \\ b_{1000} \end{pmatrix}$$

The angle of rotation was $\theta = \frac{\pi}{3}$, in other words 60°, so every six rotations would be 360°, bringing me back to where I'd started. This meant there was no need to get freaked out about having to apply it 1000 times; all I had to do was divide 1000 by 6 and find the remainder—an application of modular arithmetic—which would be 4.

$$\begin{aligned}\begin{pmatrix} a_{1000} \\ b_{1000} \end{pmatrix} &= \begin{pmatrix} a_{1000 \bmod 6} \\ b_{1000 \bmod 6} \end{pmatrix} \\ &= \begin{pmatrix} a_4 \\ b_4 \end{pmatrix}\end{aligned}$$

Four rotations by θ is the same as one rotation by 4θ, so all I needed to do was rotate point $\binom{a_0}{b_0} = \binom{1}{0}$ by 4θ.

$$\begin{aligned}\begin{pmatrix} a_{1000} \\ b_{1000} \end{pmatrix} &= \begin{pmatrix} \cos 4\theta & -\sin 4\theta \\ \sin 4\theta & \cos 4\theta \end{pmatrix} \begin{pmatrix} a_0 \\ b_0 \end{pmatrix} \\ &= \begin{pmatrix} a_0 \cos 4\theta - b_0 \sin 4\theta \\ a_0 \sin 4\theta + b_0 \cos 4\theta \end{pmatrix} \\ &= \begin{pmatrix} 1 \cdot \cos 4\theta - 0 \cdot \sin 4\theta \\ 1 \cdot \sin 4\theta + 0 \cdot \cos 4\theta \end{pmatrix} && \text{from } a_0 = 1, b_0 = 0 \\ &= \begin{pmatrix} \cos 4\theta \\ \sin 4\theta \end{pmatrix}\end{aligned}$$

Got it! The answer was $(\cos 4\theta, \sin 4\theta)$.

Next!

6.1.3 A Discovery

A bell rang, signaling the end of the test.

I always felt an indescribable sense of relief when handing in a test I'd done my best on. Math tests in particular. Math was an area I felt comfortable in, so I rarely felt like I'd done a really bad job. To use Tetra's expression, I was "friends" with math.

For my tests that day, I wasn't sure how I'd done on the other subjects, but I was pretty sure I'd gotten a perfect score on the math section. So as I headed toward the train station from my prep school, where the mock exam had been held, I was in a fine mood. The wind was a little cold, but after having worked my head for so long it actually felt refreshing.

I reflected on how important it is to perceive the form of a problem you're working on. That first problem on my math test was a prime example. The key to solving it was in seeing how it represented revolutions. This reminded me of Miruka, and a conversation we'd had long ago. I had failed to see how sine waves were reflections of rotations. Narrow-minded, she'd called me. She and I had thought through and solved many math problems in our time together.

That first problem sort of reset itself every six cycles, which reminded me of something else we'd talked about—the group axioms and the cyclic group C_6.

Definition of groups (group axioms)

A set G satisfying the following axioms is called a *group*:

- Closed under an operation $\star$.
- Associativity holds for all elements.
- Existence of an identity element.
- Existence of an inverse element for all elements.

It's easy to see that the set of all rotation matrices forms a group with respect to products of matrices. It's closed under matrix multiplication, and associativity holds for products of matrices. It has an identity element, namely $\left(\begin{smallmatrix}1&0\\0&1\end{smallmatrix}\right)$, the rotation matrix for $\theta = 0$. And the inverse element would just be the inverse of a matrix; the inverse of a rotation matrix is its reverse rotation, which would bring you back to where you started. The product of a matrix for rotation by

θ and a matrix for rotation by $-\theta$ gives the identity matrix.

$$\begin{aligned}
&\begin{pmatrix} \cos\theta & -\sin\theta \\ \sin\theta & \cos\theta \end{pmatrix} \begin{pmatrix} \cos(-\theta) & -\sin(-\theta) \\ \sin(-\theta) & \cos(-\theta) \end{pmatrix} \\
&= \begin{pmatrix} \cos\theta & -\sin\theta \\ \sin\theta & \cos\theta \end{pmatrix} \begin{pmatrix} \cos\theta & \sin\theta \\ -\sin\theta & \cos\theta \end{pmatrix} \\
&= \begin{pmatrix} \cos^2\theta + \sin^2\theta & \cos\theta\sin\theta - \sin\theta\cos\theta \\ \sin\theta\cos\theta - \cos\theta\sin\theta & \sin^2\theta + \cos^2\theta \end{pmatrix} \\
&= \begin{pmatrix} 1 & 0 \\ 0 & 1 \end{pmatrix}
\end{aligned}$$

In other words, the rotation matrix for $-\theta$ is indeed the inverse of the rotation matrix for θ.

$$\begin{pmatrix} \cos\theta & -\sin\theta \\ \sin\theta & \cos\theta \end{pmatrix}^{-1} = \begin{pmatrix} \cos(-\theta) & -\sin(-\theta) \\ \sin(-\theta) & \cos(-\theta) \end{pmatrix}$$

"Q.E.D.," Miruka might have said. "The set of all rotation matrices forms a group with respect to products of matrices."

6.2 Using Groups to Capture Forms

6.2.1 Clues in Numbers

When I arrived home, there was a surprise waiting for me—Miruka, sitting in my living room.

"Welcome home," she said. "Took you long enough."

My mother came in and started to refill the teacup on the table in front of her, but Miruka held up a hand and refused.

"I'm good, thanks."

"A piece of cake, maybe?" my mother offered.

I wondered how long it had been since Miruka had come to my house. Her presence made the place feel so very different. Not tense, exactly, but something along those lines.

"We don't need cake," I said. "What's up, Miruka?"

"We were having a nice little chat," Mom said.

"Just came to see how your mother's doing," Miruka said. "And to see Yuri if I could, but I guess I missed her. How'd the test go?"

"Okay, I guess. Pretty sure I aced the math section."

Miruka gave a slight nod. "I would hope so."

"I think I'm becoming a lot more flexible in my thinking. There was this one problem, a recurrence formula for a $\frac{\pi}{3}$ rotation matrix... Yeah, that was a good one."

"Looks like it put you in a good mood, if nothing else."

"Solving math problems always does," my mother butted in as she entered from the kitchen, carrying plates of cake. She seemed to be in a good mood herself.

"C'mon, Mom..."

"Yes, yes. I'll get out of your hair." She retreated back to the kitchen.

"So I hope rotation matrices reminded you of cyclic groups?" Miruka said.

"Of course," I said. "Speaking of which, didn't you say something the other day about groups being tools for seeing forms?"

"For investigating, knowing, and categorizing forms, yes."

"Are you talking about triangles and rectangles and things like that?" Mom said, coming back in with a cup of tea for me on a tray.

"That's exactly what I'm talking about! Sometimes numbers are good tools for categorizing shapes. For example, we can use numbers of vertices to categorize polygons."

Miruka grabbed a nearby pad of paper and made a sketch.

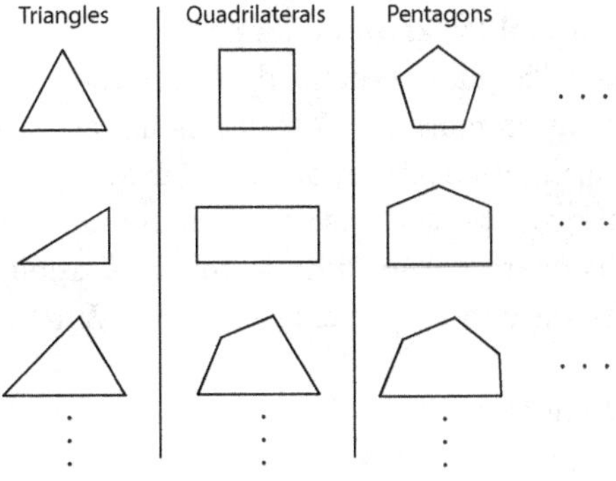

"That makes sense," my mother said.

"Mom, why don't you go—"

She grimaced. "Miruka's so nice to me, yet my own son is so cold..."

"Using numbers of vertices is a natural way to categorize polygons," Miruka continued. "In fact, we often call these shapes n-gons, as a way to consider all polygons having n vertices as being the same. So that's a categorization according to number of vertices, but that's not the only way we can do it. For example, we can categorize triangles as being acute, right, or oblique. To do so, we use the size of the triangle's largest angle as our basis."

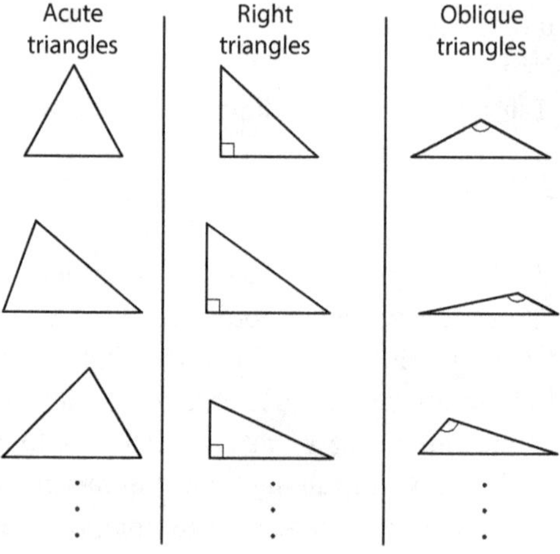

"Sure, that's fine and all, but what does this have to do with using groups as a tool for knowing forms?"

Miruka sighed. "So narrow-minded. You're the one who was just talking about rotation matrices. A $\frac{\pi}{3}$ rotation matrix does an excellent job of capturing the features of a regular hexagon. Let a be the operation 'rotate counterclockwise by $\frac{\pi}{3}$ about the origin,' and use products of that operation to indicate repeated applications. That lets us create the group $\langle a \rangle$ generated by a. What's the identity element e?"

"A non-rotating element, so $e = a^0$."

"And what's a's inverse element a^{-1}?"

"An operation saying we rotate counterclockwise by $-\frac{\pi}{3}$ about the origin, so that $aa^{-1} = e$. And the associative law holds. So if n is an integer, the set of all a^n forms a group."

"A group with what order?" Miruka pressed, making me feel as if I were taking an oral exam.

"That's the number of elements in the set forming the group, right? So that would have to be 6, the number of ways to rotate a regular hexagon."

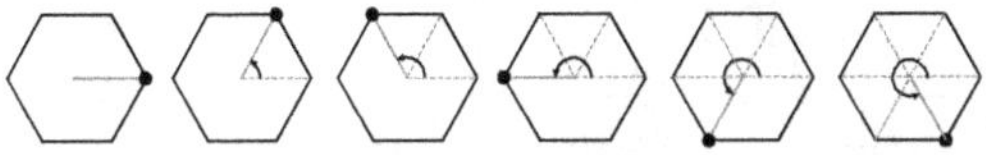

$e = a^0 \quad a^1 \quad a^2 \quad a^3 \quad a^4 \quad a^5 = a^{-1}$

"Very good," Miruka said. "We can write this group like this..."

$$\{e, a, a^2, a^3, a^4, a^5\}$$

"...or like this..."

$$\{a^0, a^1, a^2, a^3, a^4, a^5\}$$

"...or like this."

$$\{a^{n \bmod 6} \mid n \text{ is an integer}\}$$

"I know all this," I said. "That's why I was able to solve that problem on my test."

"A little review won't hurt," Miruka chided. "So six revolutions by $\frac{\pi}{3}$ is equivalent to a^6, which in turn is equivalent to a^0. A group for which all elements are generated from one element a is called a cyclic group, which I'll write as $\langle a \rangle$. The group $\langle a \rangle$ we're talking about here has order 6, so it's isomorphic to the cyclic group C_6."

"We're sort of categorizing by the number 6 here, aren't we? Even before we bring groups into the picture, I mean, in the sense that we come all the way around after six operations."

"That 'sense of coming all the way around' is one aspect of this form. Cyclic groups mathematically express that sense."

"And that's why cyclic groups are the tool allowing us to investigate forms?"

"*A* tool, not *the* tool. Cyclic groups are one of the simplest kind of groups. We can also use groups to represent more complex operations. For example, how about an operation that shows not just rotations but also reflections? For that I guess we need to show arrows in opposite directions for pairs of vertices."

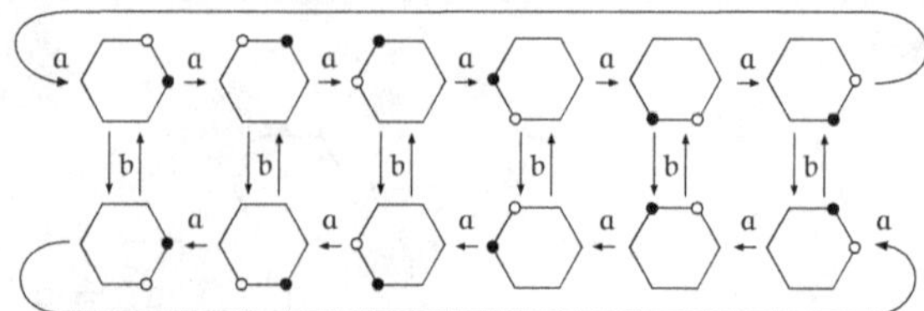

"Interesting. So the operations a for a rotation and b for a reflection can form a group too. The cyclic group C_6 is the group that is produced by a, and C_2 is the group that is produced by b, so combining the two..."

"What we see in this figure is all the patterns that the operations can produce. We used rotation and reflection operations to investigate regular hexagons. Generally, this is called a dihedral group. It's a group created from regular polygons where we distinguish between heads and tails, so to speak. Counting vertices is fine, looking at the size of angles is fine... But by using groups we can capture and represent far more complex forms. That's why I say groups are an excellent tool for learning about and categorizing forms."

"Gotcha."

"Rotations and reflections are geometric descriptions, but groups allow us to use algebraic representations. We can even use groups to describe forms we couldn't actually draw. In algebraic topology, for example, we can use algebraic methods to study topological spaces. Groups play a vital role there too."

"Using groups to study topological spaces? Isn't there something strange about that? Like, when we rotated and reflected these hexagons, their side lengths stayed the same. But doesn't topology do away with the concept of length? So I don't see how we can use groups there."

"We can always create groups, so long as the group axioms are satisfied. Considering group operations as representing rotations and reflections is just one example of how we can use them. In algebraic topology we create different kinds of groups. Topology is concerned with topological invariance, so clearly it heads in a different direction from what we've done here. Specifically, you would want to create groups that are invariant under homeomorphisms." Miruka paused, and with an amused look asked, "What do you think such a group would show you?"

"It would show you, uh... I guess I'm not really sure."

"Say you have two polygons, and you count the vertices in each. If you came up with different numbers, they couldn't be the same kind of polygon, right?"

"Well of course not."

"Same thing here. Start with two topological spaces and find groups for each that are invariant under homeomorphisms. If those groups aren't isomorphic, then the topological spaces can't be homeomorphic. So to borrow Tetra's phrase, groups are a good weapon for distinguishing between topological spaces."

"Huh."

"So now the question becomes, what kind of group might we use for topological spaces? What kind of group would be useful for explorations of the form of topological spaces? Through use of what kind of group can we move what kinds of problems from the realm of topology to the realm of algebra? That's algebraic topology, and it provides us with countless topics for research."

By this point the space in which we were sitting had disappeared. I was no longer conscious of my house, of my living room. My entire mind was occupied by Miruka's lecture.

6.2.2 Searching for Clues

When it got dark I walked Miruka to the train station.

"Your mother is so nice," she said.

"Because she thinks you're the bee's knees," I replied.

"Hmph."

Mom had caused quite a commotion before we'd left, trying to get Miruka to stay for dinner. Honestly I would have loved for Miruka to join us, but I wasn't quite brave enough to tell her what to do. So here we were, slowly walking along the path to the station.

"So was it Galois who first came up with the idea of groups?" I asked.

"I think the seeds were planted long before his time," she said, her voice low. "There are groups hidden everywhere, after all. In symmetric shapes, in the discovery of patterns, in periodic motion, in musical rhythm... Galois shined his light there, and brought these phenomena onto a mathematical stage. He studied coefficient fields as a way of determining whether equations can be solved algebraically, along with corresponding groups for investigating fields."

"Right." I nodded, recalling our adventures at the Galois Festival.[1]

"Galois described the task of finding solutions to equations as 'more obscure and perhaps more isolated' than any other problem in pure analysis. Indeed, for mathematicians of his time, the group theory he produced was something completely new. Today, however, groups are a fundamental tool in mathematics. They're indispensable for describing symmetries and other relations between mathematical objects. In that sense, Galois's words no longer hold true. Groups certainly aren't isolated from other problems. In fact, they're inherently related to many of them."

"I agree with you there," I said, "but he probably just wanted to say groups were 'isolated' from problems at the time. I think he was looking at those problems at a different level from his contemporaries. He was seeing connections that no one else was seeing."

I saw a gleam in Miruka's eyes.

"Possibly so," she said.

"So just like Galois studied groups as a way of investigating fields," I continued, "in algebraic topology I guess we study groups as a way of investigating topological spaces, yeah? This happens all the time in math, since mathematicians love building bridges between worlds."

We crossed the pedestrian bridge in front of the station, a river of cars flowing beneath us. Miruka stopped halfway across and turned to face me.

---

[1]See *Math Girls 5: Galois Theory*.

"You're right," she said. "Mathematicians love building bridges, sure. But that's not all. That's not enough. After all, what's the point of building a bridge if you don't cross it?"

I gulped and nodded. So many times I had done the same thing. I'd come this far, so of course I wanted to make that one additional step.

"Sure, I want to take that last step," I said, putting my thoughts into words.

Miruka moved closer and extended a hand. Her fingers touched my cheek, tracing a figure on it. *So warm.*

"Here is the form that is you," she said. "We can trace along you in an attempt to capture that form. If we want to know the shape of a topological space, how should we trace along it?"

I remained silent. Miruka gave my cheek a fierce twist.

"Ouch!"

She repeated her question. "How should we trace along a topological space?"

"I don't know. How?"

"Let's try creating loops," she said. "Tomorrow. In the library."

With that she glided over the remainder of the bridge and merged into the crowd in front of the station, leaving me standing there, my cheek throbbing.

6.3 Using Loops to Capture Form

6.3.1 Loops

Miruka, Tetra, and I gathered in the school library after our classes. Miruka began as soon as we'd settled in.

"Let's talk about the fundamental group, one kind of group we can add to topological spaces. The fundamental group allows us to trace along a topological space."

"Trace along it how?" Tetra asked, tracing out shapes on the desk with both hands.

"Not with a hand," Miruka said, "with a finger. Imagine tracing a finger along a torus or a sphere to examine its shape."

Miruka lightly ran her index finger around her forearm.

"We're still talking math here, right?" I said.

Miruka ignored me. "The first step toward creating the fundamental group is to create a loop in the topological space."

"Loop theory?" Tetra said, forming a shape with her thumbs and index fingers that looked to me more like a heart.

"A loop like this." Miruka spun a finger in a circle. "Let some point in the topological space be your starting point. From there, draw a path through the space, ending where you started. That kind of loop. Since the initial and terminal points for this path are the same, let's call that the base point."

"So we're looping back around to where we started. Got it."

"Drawing a curve in a topological space means we've continuously joined elements, in other words points, belonging to that space. Continuously joining two things requires the concept of 'nearness,' but we're in a topological space so that isn't a problem, since we can define nearness using open neighborhoods."

Tetra raised a hand. "Sorry, but could you give me an example?"

"Sure. Think of a point p fixed on the surface of a donut, in other words a torus. If p is our base point, we can create a loop like this."

"Just a plain old loop!" Tetra said.

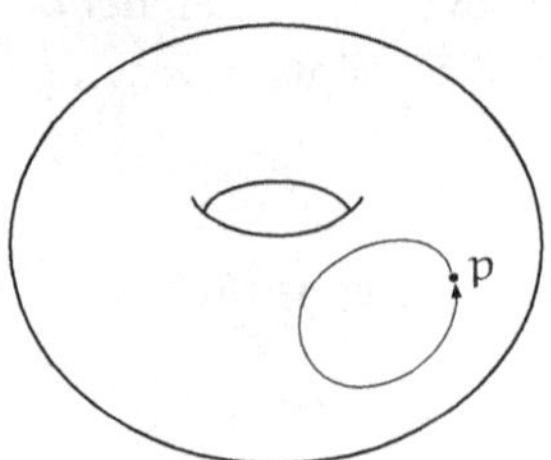

A "plain loop" on a torus.

"Huh. I was imagining something like this," I said.

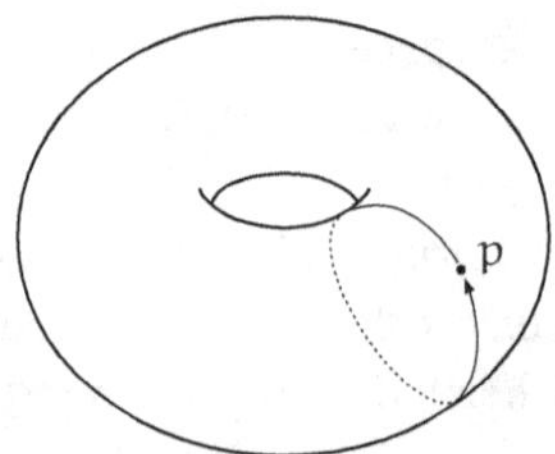

A "small loop" on a torus.

"Oh, a small loop through the donut hole!"

"I guess we could also make a big loop, like this." I made a third drawing.

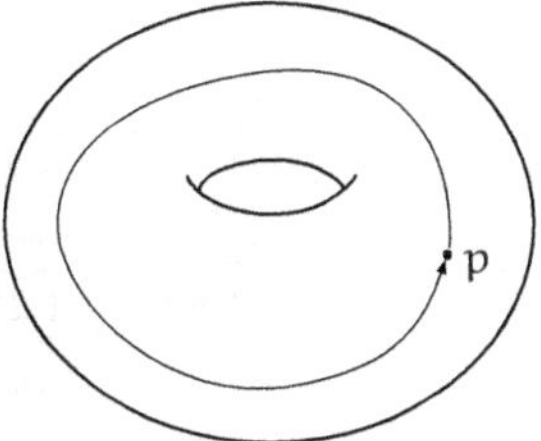

A "big loop" on a torus.

"Well done," Miruka said. "Each of these are loops on the topological space of a torus. Each starts and ends at the same point, none of them are broken anywhere, and none of them leave the surface of the torus. So yes, each one is a loop."

"Okay, I think I'm good with loops," Tetra said, nodding.

"Good. Then let's try representing them mathematically. Consider the closed interval $[0, 1]$. This is the set of all real numbers t such that $0 \leqslant t \leqslant 1$. Now consider a continuous mapping f from the closed interval $[0, 1]$ to the topological space. In other words, for every real number t from 0 to 1, $f(t)$ represents one point in the topological space. Further, the mapping f satisfies the condition $f(0) = f(1)$. This is the loop mathematically represented by the continuous mapping f."

"Uh, hang on. Where did this $f(0) = f(1)$ condition come from?" Tetra asked.

"I think it shows that our initial and terminal points are the same," I said. "So $f(0) = f(1) = p$."

Miruka nodded. "That's right. Here's how it works for the big loop."

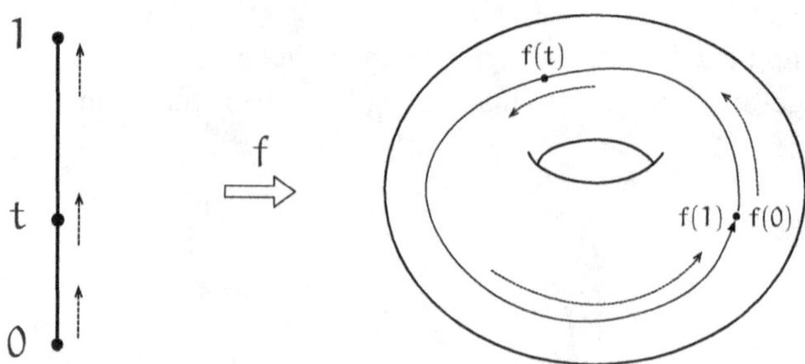

Representing the big loop as a continuous mapping f.

"Ah, okay," Tetra said. "So $f(0)$ is the initial point, and $f(1)$ is the terminal point. When we move t from 0 to 1, the point $f(t)$ moves across the torus, just like we're tracing a finger across it."

Tetra made a circle in the air in imitation of Miruka.

"Except it's the point $f(t)$ that's doing the tracing," I said.

"There are infinitely many ways to make a loop on a torus," Miruka said. "All we have to do is bend a loop a tiny bit somewhere along its path to turn it into a different loop. So we'd like some way to consider all continuously deformed loops as being the same. Some method for handling all continuously deformed loops in the same way. We've represented loops as a continuous mapping, so continuously deforming these loops is equivalent to continuously deforming our continuous mapping."

"That's going to require some explanation."

"Happy to oblige," Miruka said.

6.3.2 Homotopic Paths

"We can imagine a loop in a topological space as something like a rubber band wrapped around a torus," Miruka said, "and we can think of a continuous deformation of that loop as pulling the rubber band into a new tracing. This is exactly the same as thinking of a continuous mapping H from the Cartesian product $[0,1] \times [0,1]$ to the topological space. For example, We can use H to deform a loop f_0 on the torus to another loop f_1, keeping the same base point."

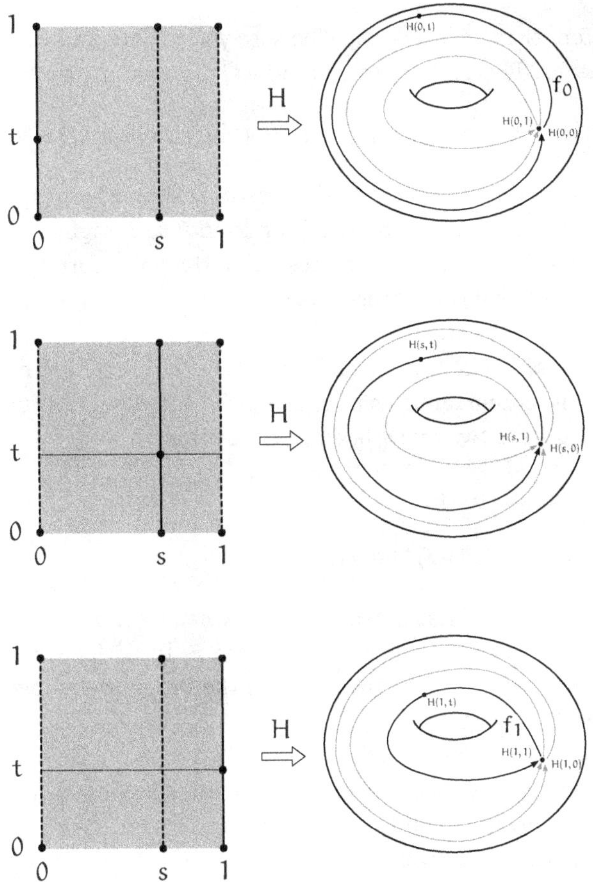

Tetra and I sat for a moment, looking at the graphs Miruka had sketched.

"Okay, sure," I said. "Now I see what you mean by continuously deforming a loop represented as a continuous mapping. So this is how I should have been imagining all those droopy forms I see in topology books. I should think of 'connected' as implying the existence of a continuous mapping, and 'deforming connected things without cutting them' as a continuous deformation of that continuous mapping."

"I'm sorry, I'm still not sure what this continuous mapping H is," Tetra said.

"Before that," Miruka said, "be sure you understand what $[0,1] \times [0,1]$ means. This is a Cartesian product, representing a set like this."

$$[0,1] \times [0,1] = \{(s,t) \mid s \in [0,1], t \in [0,1]\}$$

"This would be the set of all pairs of real numbers (s,t), where s and t are both able to be anything from 0 to 1, right?"

"That's right. And we're representing the point corresponding to (s,t) in the topological space as $H(s,t)$."

"I'm still not sure—"

"Think of it like this," I said. "You have $H(s,t)$. First fix s, then move t with s remaining as it is, and make sure $H(s,t)$ forms a loop. Then, you have two loops f_0 and f_1, like this."

- When $s = 0$, $H(0,t) = f_0(t)$, and
- when $s = 1$, $H(1,t) = f_1(t)$.

"So that's how $H(s,t)$ behaves. Now when we move s from 0 to 1, we create a continuous deformation from f_0 to f_1. This is the same idea as moving t to move the point on the torus. We're just moving s instead to deform the loop."

"So the loop is deformed, but it's still a loop?"

Miruka nodded. "Of course. We're fixing the base point p here, so we can add the condition that for any H, $H(s,0) = H(s,1) = p$ for all values of s from 0 to 1."

"Okay, let me make sure I've got this straight. We move t with $H(0,t)$ to create this one loop f_0, then we move t with $H(1,t)$ to create another loop, f_1. Then we move s with $H(s,t)$ to move from f_0 to reach f_1?"

"You've got it. If you have a topological space—a torus, say—you can create infinitely many loops on it. In other words, there are infinitely many ways to trace a finger on it. We can categorize those infinitely many loops according to whether they can be continuously deformed to coincide."

"Interesting. Another way to see equivalencies. In this case, equivalency through stretching and shrinking, in other words the existence of a continuous mapping H."

"That's exactly right. When we fix the base point p, a continuous mapping H that allows us to transition between loops f_0 and f_1 is called a homotopy. When a homotopy H exists, we say f_0 and f_1 are homotopic loops, meaning we can continuously deform f_0 into f_1. We show that like this."

$$f_0 \sim f_1$$

"This is an equivalence relation, so we can divide the set F of all loops having base point p by this relation, giving homotopy classes as collections of equivalent loops."

6.3.3 Homotopy Classes

"So we have a topological space," Tetra said, speaking to herself to organize her thoughts, "and we have loops in that topological space, but there's lots of loops we could make, so we're considering all loops that can be continuously deformed into each other as being the same?"

Miruka nodded. "That's right."

"We still haven't talked about fundamental groups, right? Or did I just miss all that?"

"No, we haven't talked about those yet. We're still building up the elements we need to create them."

"And loops are one of those elements?"

"Not loops themselves. The element we're after is 'collections of loops that can be considered equivalent through continuous deformations.' We can divide the set of all loops by the equivalence relation of homotopic loops, giving us some number of homotopy classes."

"Sorry, can you give me an example? Like, on a torus?"

"Sure, let's think about which loops on a torus would be homotopic, and gather them up into sets. Each set will be a homotopy class. For example, the 'big loop' homotopy class would be the set of all big loops passing through point p."

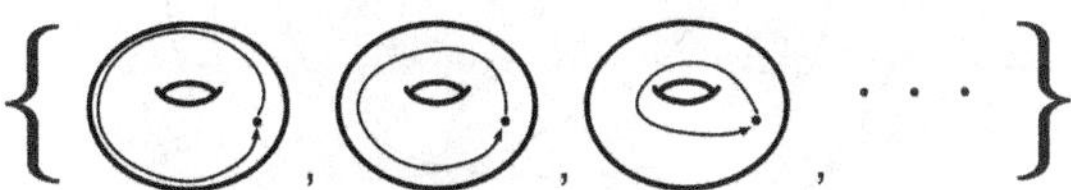

The "big loop" homotopy class.

"Okay, we want to make a set of all the loops that can continuously change into each other. Got it. So I guess the homotopy class for the small loops would be something like this?"

The "small loop" homotopy class.

"And this would be the 'plain loop' homotopy class," I said.

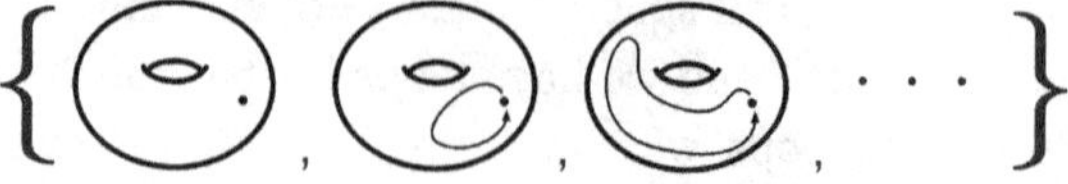

The "plain loop" homotopy class.

"Hang on," Tetra said. "That one on the left isn't a loop, is it?"

"Sure it is. A loop made from a single point. It satisfies $f(0) = f(1)$, after all, and there are no rules saying we have to move to a different point in-between."

"I guess you're right. I need to be careful of my preconceived notions about what a loop is. So anyway, there's three homotopy classes for a torus?"

"Looks like it."

"Wrong," Miruka said, shaking her head. "There are infinitely many homotopy classes on a torus. Here's one you missed."

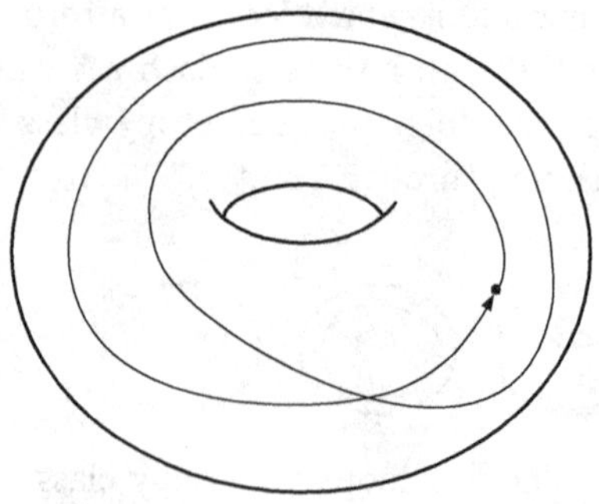

Tetra's eyes went wide. "Oh! A double loop!"

"Aha," I said. "Because there's no way to continuously turn a double loop on a torus into a single loop."

"Or how about this?" Miruka said, drawing another sketch.

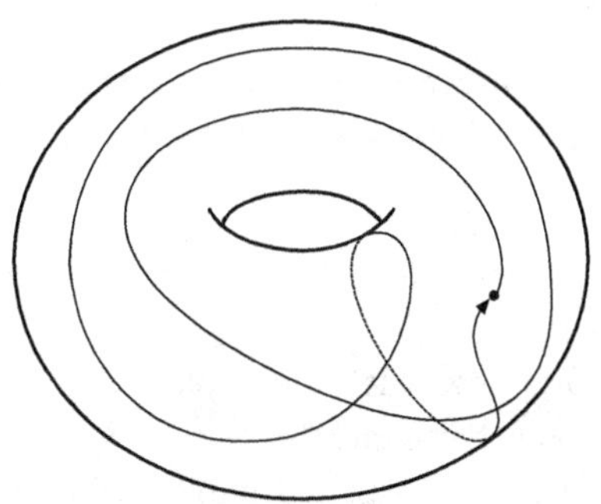

"Ooh, that's a good one too!" Tetra said.

"Hang on, let me— Ah, okay. So a loop formed by connecting a big loop and a small loop produces a new homotopy class. Got it."

"So this means we can make all kinds of patterns of loops and wraps and their combinations!"

"Exactly," Miruka said. "That's the key idea behind homotopy groups."

6.3.4 Homotopy Groups

"Let's consider joining a loop as a group operation," Miruka began. "In turn, it goes something like this."

- Fix a point in a topological space X, and consider it as the base point p.
- Let F be the set of all loops having base point p.
- Divide set F by the loop homotopy equivalence relation $\sim$ to obtain the set $F/\sim$ of all homotopy classes.
- Elements of set $F/\sim$ are sets of loops that can be considered equivalent as continuously deformable loops.

- Add a group operation to make set $F/\sim$ a group. Namely, we consider joining elements of set $F/\sim$, in other words homotopy classes, as an operation, and set this as the group operation. Recall that all loops share the base point p, so they can always be connected.

- We call a group created in this way the "fundamental group for the topological space X with base point p," and write this as $\pi_1(X, p)$.

"This pi here," Tetra said. "That doesn't have anything to do with circles, does it?"

Miruka shook her head. "It's a completely different thing. The π in $\pi_1(X, p)$ is just another character."

"Gotcha. It just kinda surprised me, showing up here."

"It isn't difficult to confirm that $\pi_1(X, p)$ satisfies the group axioms," Miruka continued. "For example, what would be the identity element?" She pointed at Tetra.

"That would be a loop that doesn't change when it gets joined, so I guess that's the plain loop?"

"Actually, the homotopy class for the plain loop," I added.

"That's right," Miruka said. "Specifically, the identity element is the homotopy class for loops formed only from point p."

"So intuitively speaking, that would be the set of all loops that can be continuously deformed down to the single point p, right?"

Miruka nodded. "Well put. Until now we've been fixing the base point p to create our joining operator, but actually we can think of the fundamental group without doing that. If we consider a different base point q, we can just consider curves passing between p and q. But when we do so, we have to add a condition saying that we can connect any two points in the topological space X, in other words that the space is path-connected. If X is path-connected, we can just write $\pi_1(X)$ instead of $\pi_1(X, p)$, without explicitly specifying a point p. So $\pi_1(X)$ means 'the fundamental group for a path-connected topological space X.'"

"Miruka—" Tetra groaned.

"We can also show that the fundamental group is topologically invariant. In other words, we can show that if two path-connected topological spaces X and Y are homeomorphic, then their fundamental groups $\pi_1(X)$ and $\pi_1(Y)$ are isomorphic."

"Miruka, my head is—"

"The fundamental group for a torus is the Cartesian product $\mathbb{Z}\times\mathbb{Z}$ of two additive groups $\mathbb{Z}$. These two additive groups will respectively correspond to the number of times you've wrapped around as a large loop and a small loop." Miruka paused for a moment of thought, then continued: "Let's think of a simpler topological space than a torus. What would be the fundamental group $\pi_1(S^1)$ for the 1-sphere S^1?"

Problem 6-2 (the fundamental group for S^1)

Find the fundamental group $\pi_1(S^1)$ for the 1-sphere S^1.

Just as she posed this problem, however, Ms. Mizutani appeared to make her customary announcement.

"The library is *closed*."

6.4 Capturing Spheres

6.4.1 At Home

I'd settled into a regular daily schedule: studying for exams until midnight, then jumping in the shower. I tried to stick to that timetable as closely as possible, doing the same thing at the same time every day. My plan was to become less of a night owl as my exam date approached.

Glancing at the clock, it read 23:53.

23 *and* 53. *Two primes.*

I straightened up my desk and headed for the bath. As I undressed, I recalled the day's Miruka lecture, and Tetra's response to it. And Miruka touching my cheek...

Enough of that. Focus on the problem.

6.4.2 Fundamental Group for the 1-Sphere S^1

I started thinking about Miruka's problem as I washed.

I figured I could just consider the topological space of S^1 as a circle. Then I would create loops on that circle, and consider loops that can be continuously deformed into each other as equivalent. That would mean all loops traveling about the circle and ending up at the same place, even if they did some back-and-forthing along the way, are equivalent. A loop traveling once around S^1 and one traveling twice around it would be different, though, since they can't be continuously deformed into each other. So I could categorize loops by the number of times they went around the circle. Oh, and they can go in either direction. So once any back-and-forth movements have been untangled, I could create the fundamental group for S^1 just from numbers of loops and their directions.

In the end, it looked like I was ending up with a group formed by the integers—an additive group created from the set of all integers with the addition operation + added.

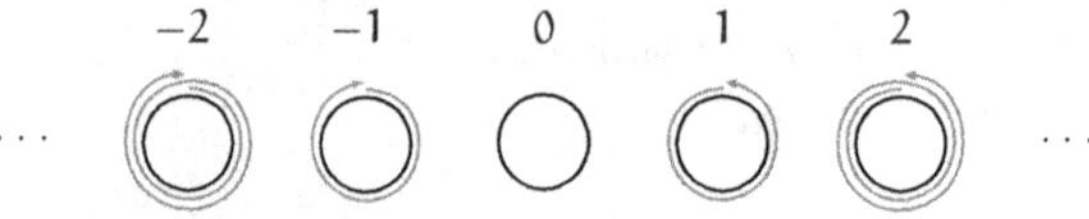

In other words, the fundamental group $\pi_1(S^1)$ for the 1-sphere is isomorphic to the additive group $\mathbb{Z}$.

$$\pi_1(S^1) \simeq \mathbb{Z}$$

The element equivalent to 0 in $\mathbb{Z}$, in other words the identity element e, would be the homotopy class for a loop created from a single point, the set of all loops that do not completely traverse the circle.

Interesting. I'm starting to get a feel for how fundamental groups can use loops to capture forms.

Sure enough, the structure behind the form of the 1-sphere S^1 did feel a lot like the structure of $\mathbb{Z}$. I could think of "traveling once around the circle, then two more times, for three laps in total" as $1 + 2 = 3$. "Taking two laps around the circle and not going back" would be $2 + 0 = 2$. "Going some number of laps in one direction, then going back the same number of laps" would be $n + (-n) = 0$.

And I could summarize all these different ways of traveling around S^1 just by saying "the fundamental group for S^1 is isomorphic to $\mathbb{Z}$."

Answer 6-2 (the fundamental group for S^1)

The fundamental group $\pi_1(S^1)$ for the 1-sphere S^1 is isomorphic to the integer additive group $\mathbb{Z}$.

6.4.3 Fundamental Group for the 2-Sphere S^2

While shampooing, I decided I'd like to take things up a notch. What would the fundamental group for the 2-sphere S^2 be?

Problem 6-3 (the fundamental group for S^2)

Find the fundamental group $\pi_1(S^2)$ for the 2-sphere S^2.

I could think of a 2-sphere as the surface of a ball, on which I'd be making loops. Those loops would have to be continuous curves with the same start and end points, and they couldn't stick out from the topological space. I would consider any loops that could be continuously deformed into each other as being equivalent.

I tried imagining what kinds of loops I could trace out on a ball. Perhaps I just wasn't imaginative enough, but it seemed like any two such loops could be continuously deformed into each other, so there seemed to be only one kind—a loop formed from a single point. Any other loop could be collapsed down to that one, after all. In other words, all loops on a 2-sphere are homotopic.

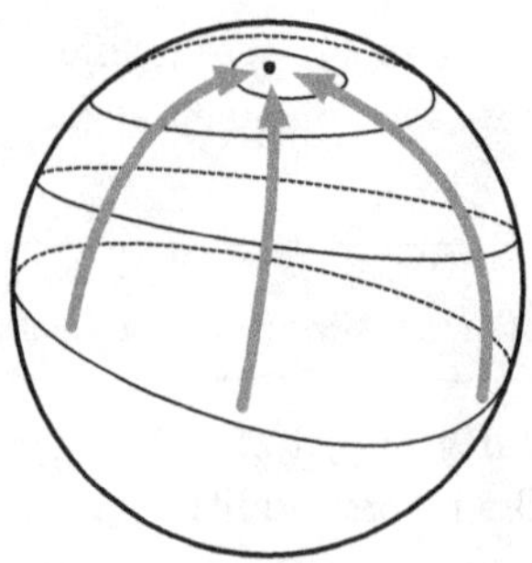

Compressing loops on a 2-sphere to a single point.

That would mean the fundamental group for a 2-sphere would have only one element, the identity element e, making it the trivial group $\{e\}$.

So unless I'm missing something, the fundamental group $\pi_1(S^2)$ for the 2-sphere S^2 is isomorphic to the trivial group $\{e\}$!

$$\pi_1(S^2) \simeq \{e\}$$

Answer 6-3 (the fundamental group for S^2)
The fundamental group $\pi_1(S^2)$ for the 2-sphere S^2 is isomorphic to the trivial group $\{e\}$.

There was a clear difference in how the 1-sphere and the 2-sphere felt. I figured it must be because of the "hole" in the 1-sphere, which loops could get wrapped around, preventing the collapse of loops down to a single point. The 2-sphere didn't have a loop-catching hole, so any kind of loop you made there was continuously collapsable down to a point.

6.4.4 Fundamental Group for the 3-sphere

Getting in the tub for a post-shower soak, I decided to go up one more dimension. An idea suddenly came to me—

Won't this one be the same?

The 3-sphere isn't like a ball, it's more like three-dimensional space. So no matter how I traced out a loop in space, shouldn't I be able to just shrink it down to a single point?

Ah! So that's *what they were talking about!*

I'd recalled the television show about the Poincaré conjecture I'd seen, where they'd shown a graphic of a spaceship tugging on a rope.

Topological space M	Fundamental group $\pi_1(M)$
1-sphere S^1 (circle)	Additive group of integers $\mathbb{Z}$
2-sphere S^2 (ball surface)	Trivial group $\{e\}$
3-sphere S^3	Trivial group $\{e\}$

That seemed right, but . . . strange. The 2-sphere and the 3-sphere have the same fundamental group? So how could you use fundamental groups to distinguish between their forms?

6.4.5 The Poincaré Conjecture

Getting out of the bath, I hurriedly dried my hair, then rushed to my bookshelf. There, I found a book on topology that I'd bought quite a while back, but had set aside as too difficult for me at the time.

That book called the fundamental group the "first homotopy group," which it used to create groups from loops. So, groups created from 1-spheres. A group created from 1-spheres is the first homotopy group $\pi_1(M)$, a.k.a. the fundamental group. Finally I realized why Miruka's notation for the fundamental group had that subscripted 1.

So I guess the group created from n*-spheres would be the* n*-th homotopy group,* $\pi_n(M)$*? A group generalizing the concept of fundamental groups! And I suppose that as a three-dimensional being, I can use the second homotopy group* $\pi_2(S^2)$ *to describe how I experience the space of* S^2*!*

I continued reading, and soon came across the Poincaré conjecture.

The Poincaré conjecture

Let M be a three-dimensional closed manifold. If the fundamental group for M is isomorphic to the trivial group, M is homeomorphic to the 3-sphere.

Finally, I had some idea of what this proposition meant.

- I knew what a three-dimensional closed manifold is. That's a three-dimensional topological space that's locally homeomorphic to three-dimensional Euclidean space and is bounded but without boundary. If I were to jump into such a manifold and look around, it would be just like floating in outer space. I'd be able to head off any distance in any direction and never reach an end. That's what it meant to be "bounded but without boundary."
- I also knew what the fundamental group for M is. That's the group $\pi_1(M)$, created by considering all loops that can be continuously deformed into each other as equivalent.
- I even knew what a 3-sphere was. That's the topological space S^3, something like two 3-balls with their surfaces connected.

So now I could understand not only the content, but the intent behind this assertion that the Poincaré conjecture was presenting—that if the fundamental group for M is isomorphic to the trivial group, M is homeomorphic to the 3-sphere. The Poincaré conjecture is an attempt to know the power of the fundamental group. If it is true, then the fundamental group becomes a tool for determining whether an object is homeomorphic with S^3.

Say you want to investigate some three-dimensional closed manifold M. What kind of manifold is it? For example, is it homeomorphic to S^3? Well, just find its fundamental group, $\pi_1(M)$. If you end up getting the trivial group, then yes, it is. Otherwise, no. So rather than comparing M and S^3 in the world of topological spaces, you could find the answer by comparing $\pi_1(M)$ and $\pi_1(S^3)$ in the world of groups. The Poincaré conjecture would be your bridge between those two worlds.

If, that is, it is true.

So while it is an effort to determine the power of fundamental groups, the Poincaré conjecture is also a problem that combines topological geometry and algebra.

6.5 Captured by Form

6.5.1 *Confirming Conditions*

"I stayed up *way* too late last night," I told Miruka and Tetra the next day in the library. "It was fun, though. I'm not sure I entirely understand everything I was studying, but I definitely had fun doing it. I think I'm finally starting to get an idea of what the Poincaré conjecture is all about."

I gave them a rundown of what I'd learned.

"So the fundamental group is like a detection tool, right?" Tetra said. "But wait, the fundamental group is a topological invariant, so it shouldn't be a surprise that two topological spaces are homeomorphic if their fundamental groups are isomorphic, should it?"

"Watch out there," Miruka said. "You have to be sure to distinguish between necessary and sufficient conditions." She opened my notebook to a fresh page and started writing. "Let $P(M)$ be the condition 'the fundamental group for M is isomorphic to the trivial group,' and let $Q(M)$ be the condition 'M is homeomorphic to the 3-sphere.' The Poincaré conjecture says this."

$$P(M) \Longrightarrow Q(M)$$

"But the topological invariance of the fundamental group tells us the opposite."

$$P(M) \Longleftarrow Q(M)$$

"Because the fundamental group is topologically invariant and the fundamental group of the 3-sphere is isomorphic to the trivial group, we can say if $Q(M)$ then $P(M)$. Tetra, are you sure you understand topological invariance?"

Tetra blinked several times. "I... I think I do, though I might be a bit fuzzy around the edges."

"Actually, I got tripped up in the same way," I said. "Saying the fundamental group is topologically invariant is asserting that if X and Y are homeomorphic, then $\pi_1(X)$ and $\pi_1(Y)$ are isomorphic. So just saying that the fundamental group is topologically invariant isn't enough. We can use that to prove that X and Y are *not* homeomorphic, but we can't use it to prove they *are*."

6.5.2 Capturing the Hidden Self

"I guess I've still got some learning to do," Tetra said. "I mean, just look at how hard you work at this stuff. By midnight, I'm well off into dreamland."

"No, no. I should have been too," I said, truthfully; I was really feeling the lack of sleep, which for some reason was making me more talkative than usual. "I wasn't even studying anything related to my entrance exams, which is what I should be focusing on. I need to be tuning myself up, like a Formula One racer. I should be taking my mock exams more seriously, to better measure how well I'll do on the real thing. Speaking of which, there was a problem on my test the other day that reminded me of cyclic groups. It involved an isomorphism with the cyclic group C_6. Not directly, mind you, but still."

Tetra's eyes went wide. "What? Cyclic groups show up on college entrance exams?"

"No, it just reminded me of them. The problem was actually about a simple rotation matrix, though it wasn't even in matrix form. It involved a rotation by $\frac{\pi}{3}$, which made me think of hexagons, so I— Hey, wait a minute... "

My heart skipped a beat.

That problem had given $\theta = \frac{\pi}{3}$, a specific rotation angle. That was exactly why I'd been able to use mod 6 to find a solution. I'd left my solution as

$$(a_{1000}, b_{1000}) = (\cos 4\theta, \sin 4\theta),$$

but I could have kept going!

$$\begin{cases} \cos 4\theta &= \cos \frac{4\pi}{3} = -\frac{1}{2} \\ \sin 4\theta &= \sin \frac{4\pi}{3} = -\frac{\sqrt{3}}{2} \end{cases},$$

so...

$$(a_{1000}, b_{1000}) = (-\tfrac{1}{2}, -\tfrac{\sqrt{3}}{2}).$$

I'd forgotten to perform the substitution! What a stupid mistake!

Answer 6-1 (a recursion formula)

$$(a_{1000}, b_{1000}) = (-\tfrac{1}{2}, -\tfrac{\sqrt{3}}{2})$$

"You okay?" Tetra asked, peering up at me in concern.

"Yeah, I'm fine. Just noticed a mistake. On my mock exam."

"Math?" Miruka asked.

"Yeah, forgot to make a substitution in an answer."

Miruka shrugged. "You'll get partial credit," she said with her usual indifference.

Her usual indifference, yes, but this time it really hit a nerve.

"How can you be so spiteful?" I said, perhaps a bit loudly.

She narrowed her eyes. "Spiteful, am I?"

"I mean, look at you. You've already been accepted at a good school. You don't have to take any exams. Yet you just sit there all smug, talking about my partial credit!"

"Smug, am I?"

"You always are, with that 'I know everything' attitude, acting so...so...ruminative!"

"Ruminative, huh?"

"Aloof!"

"Aloof."

"Detached!"

"Detached."

No, not this. This pitiful whining, this wasn't what I wanted to say.

"Is that the end of your vocab list?" Miruka said. "Well, good to know how you really feel about me, I guess."

It wasn't how I really felt about her, of course, but I couldn't find the right words. I was just spewing nonsense to cover up how ashamed I was of myself, and as a way to hold back my tears.

"So," Miruka said. "You get partial credit on one problem on one mock exam, and that's enough to leave you shocked and dumbstruck? Seriously?" She started speaking very slowly. "Dude, it was a *mock exam*. It *doesn't matter*. Your first one doesn't matter, your millionth one doesn't matter. It's *practice*."

I just sat there, silent.

"Listen," she continued. "What you see of me is not all of me. What I see of you is not all of you." She paused. "But today you certainly have shown me more of yourself."

> Consider a compact three-dimensional manifold V without boundary. Is it possible that the fundamental group of V could be trivial, even though V is not homeomorphic to the three-dimensional sphere?
>
> HENRI POINCARÉ

CHAPTER 7

The Warmth of Differential Equations

> The rate of change of an object's temperature is proportional to the difference between its own temperature and the temperature of its surroundings.
>
> NEWTON'S LAW OF COOLING

7.1 Differential Equations

7.1.1 *In the Music Room*

"Sounds like something you'd do," Ay-Ay said. "Yeah, that's all on you, dude."

We were in the music room, where Ay-Ay was playing a piece on the piano. I was standing next to her, entranced by the dance of her long fingers as she played a jazz arrangement of something by Bach.

Ay-Ay was in my grade. She was quite the pianist, though the first thing you would notice about her would be her long, wavy hair. Another student had taken over her former position as president of the school's piano club, but she was still a constant presence in the music room.

"I know, I know," I said. She was speaking, of course, about my little dust-up with Miruka from yesterday, which I'd filled her in on.

"What did you think would happen, taking out your frustrations on her? I mean, she was just speaking truth. You did partial credit work, so live up to it."

"I'm just a partial credit kinda guy, I guess."

Ay-Ay was right, of course, and I knew full well I had overreacted and taken things out on Miruka. And all over one stupid mistake. One of many.

"Music is an art that plays out over time, and time is irreversible," she said. "You can't take back a wrong note, you can only keep going. You can't stop the music."

"Ever onward, yeah."

"Time is also one-dimensional," she continued, "but directional. People have memories, and they remember what they've heard. Memories of sounds chain together, leaving patterns in their hearts. If you want to make good music, you need to chain together good sounds. As maestro says, every note is for the piece, the piece is for every note."

Ay-Ay had long tread a musical path, under the tutelage of her "maestro," a master musician, since she was little.

"I guess the problem is finding the good sounds," I said.

"Time is also continuous," she said. "In music, you can't pull out a single note and call it a 'good sound.' A good sound has to come at the right time, and the right time depends on all the sounds surrounding it." She stopped playing and turned to face me. "Maybe the same holds for the things we say, ya think?"

Right again. No point in clinging to my stupid mistakes. No point in letting my pathetic attitude ruin an important relationship.

"Thanks, Ay-Ay," I said as I headed out of the music room.

I heard her resume her Bach behind me.

7.1.2 In the Classroom

"Well you sure look glum," Tetra said. "Okay to join you for lunch?"

She had come to my classroom at lunchtime. It was uncommon for younger students to enter the seniors' domain, but that didn't hold Tetra back, and it had become far too cold for lunch on the roof.

I looked up and said, "Sure, of course."

Tetra pulled an empty desk over to mine. "I don't see Miruka. She's out today?"

"Looks like it," I said, spreading out my lunch. I was particularly aware of her absence because I'd intended to talk to her about—honestly, to apologize for—what I'd said the day before. Without her being there, I couldn't help but feel I'd missed any opportunity to say the right thing at the right time.

We ate in silence, which obviously made Tetra uncomfortable. After finishing her lunch, she sat there fidgeting.

"You, uh, working on anything interesting?" I asked to break the silence.

"Well, kind of" she replied, clearly relieved. "Maybe you can help. What can you tell me about differential equations?"

"Differential equations? Wow, that's a deep dive."

"No, no. I've barely stuck a toe in. It's just something that came up when I was talking to Lisa the other day. I looked it up in a book in the library, but I couldn't make heads or tails of it. I was wondering if maybe . . . ?"

"You bet," I said, pulling some paper from my desk. "Say you know $f(x)$ is a differentiable function whose inputs and outputs are in the set of all real numbers, but other than that you don't specifically know what kind of function it is. One thing you do know, however, is that this holds for any value of x."

$$f'(x) = 2$$

"This is just an example, mind you," I said. "But anyway, if you know the derivative of $f(x)$ will always equal 2, what can you say about that function?"

"Huh, lemme think. I'm trying to find $f(x)$, given that I know $f'(x) = 2$, right? I guess that means the slope of $y = f(x)$ is always 2, which means we're dealing with a linear function. So I guess the function is this?"

$$f(x) = 2x$$

"Because the graph for $y = 2x$ is a straight line with slope 2!"

"Sure, okay. You're right that the derivative of $f(x) = 2x$ would be $f'(x) = 2$—"

"Oh, good!"

"—but that's not the only one. For example, how about this?"

$$f(x) = 2x + 3$$

"Oh, of course. Because $f'(x) = (2x+3)' = 2$, so sure, that works too. As would $f(x) = 2x + 1$, for that matter."

"It would. More generally, any $f(x)$ in this form will work."

$$f(x) = 2x + C \qquad \text{for constant } C$$

"Right!"

"You thought about the slope of $y = f(x)$, which is a good approach, but another would be to integrate both sides of $f'(x) = 2$, giving us the same thing."

$$f(x) = 2x + C \qquad \text{for constant } C$$

"It does, doesn't it!"

"So this example of $f'(x) = 2$ I just gave you is a differential equation for a function $f(x)$."

$$f'(x) = 2 \qquad \text{example differential equation for } f(x)$$

"That's it? *This* is a differential equation?!"

"Well, a very simple one, but sure."

"I see. I guess I can see the 'differential' part of this, but it isn't much of an equation..."

I smiled. "It doesn't have to be complex to be an equation. Like, this is an equation, right?"

$$x^2 = 9 \qquad \text{example equation in } x$$

"The x here stands for some number, or numbers, but we don't know what it, or they, might be. We don't know what kind of number x is. That doesn't mean we don't know anything, though. We know its square must equal 9, for example. In this case, we call finding any values for x that satisfy this equation $x^2 = 9$ 'solving the equation.'"

"I'm good with all that," Tetra said.

"Of course. That's the basics of *equations*, and the basics of *differential* equations is pretty much the same. The main difference is that when we solve equations we're looking for numbers, but when we solve differential equations we're looking for functions. In the example we were looking at, we don't know what kind of function $f(x)$ is, but we do know that it satisfies the equation $f'(x) = 2$. And of course, finding that function is called 'solving the differential equation.' Makes sense so far?"

"It does. Actually, I'm pretty sure the book I was reading said more or less the same thing, but in a much more difficult to understand way. Anyway, sure, I understand what a differential equation is now."

	Example	What we're looking for
Equation	$x^2 = 9$	Number or numbers x
Differential equation	$f'(x) = 2$	Function or functions $f(x)$

"You can solve the equation $x^2 = 9$, right?"

"Well of course. The solution is $x = \pm 3$."

"Right, there are two possible solutions to $x^2 = 9$, either $x = 3$ or $x = -3$, since both of these make the equation true. So there isn't necessarily just one solution to any equation, and 'solving the equation' usually means finding *all* possible solutions."

"Yep, still good with all that."

"Well, the same thing holds for the differential equation $f'(x) = 2$. You found the solution $f(x) = 2x$, but that's not the *only* solution. $f(x) = 2x+1$ works too, and so does $f(x) = 2x+5$, and so does $f(x) = 2x - 10000$. To generalize that, to represent *all* possible solutions, we have to use the form $f(x) = 2x + C$."

"Okay, sure. You're right, equations and differential equations are similar."

"A single function like $f(x) = 2x + 1$ satisfying a differential equation is called a particular solution. In contrast, a function like $f(x) = 2x + C$ using a constant C to represent all possible solutions to a differential equation is called a general solution."

"Hmm, so doesn't that mean a differential equation like $f'(x) = 2$ will have infinitely many solutions, since C can be any real number?"

"That's exactly right. It's pretty common for a differential equation to have infinitely many solutions."

"Got it."

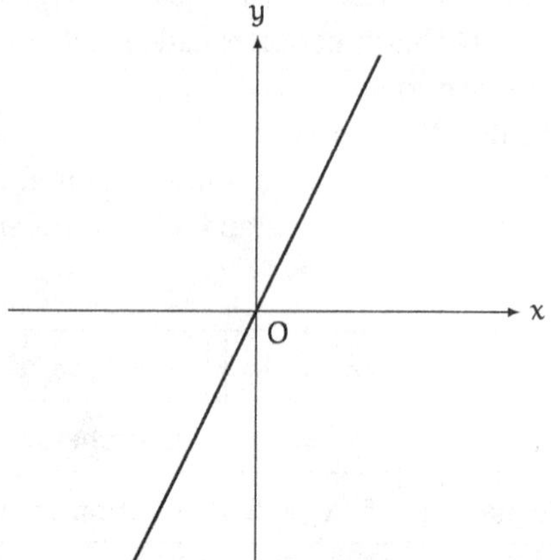

A graph of the particular solution $y = 2x$

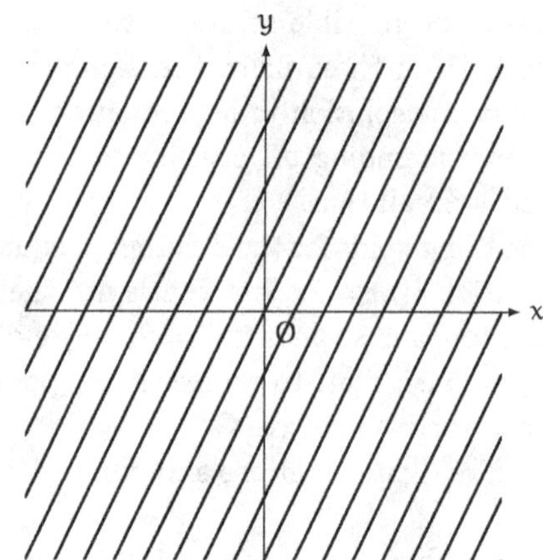

Sample graphs of the general solution $y = 2x + C$

"It's easy to check a solution you've found for the equation $x^2 = 9$. Just replace the x with your solution and make sure the equation holds. So while it can be hard to find solutions to an equation, checking a solution you've found is simple. Similarly—"

"Hold up," Tetra said, raising a hand. "You're going to say the same thing about differential equations, right? That if we want to check a solution to the differential equation $f'(x) = 2$, we just have to differentiate it. And if differentiation gives us a 2, we've found a solution. Right?"

"Exactly right. When you've found a function $f(x)$ as a candidate solution for satisfying the differential equation, you can just differentiate it to make sure it really is a solution."

7.1.3 *Exponential Functions*

Tetra was a very good listener, which made me want to speak all the more.

"So we've found this solution $f(x) = 2x + C$ for $f'(x) = 2$. More generally, we can consider calculation of the antiderivative as solving a simple differential equation. Antiderivatives produce constants of integration, which become parameters in our general solution. Assigning that parameter a specific value gives you a particular solution to the differential equation."

"Which means we can solve a differential equation just by taking the integral of both sides, right?" Tetra asked. "So if I see a differential equation, I can just integrate to whip up a solution!"

"Not so fast now. That works for the simplest possible examples, like $f'(x) = 2$, but it won't always."

"Darn."

"Differential equations can contain combinations of $f(x)$ and $f'(x)$ and $f''(x)$ and so on, so finding a solution might involve a lot more than simple integration. For example, how about this as a slightly more complex example?"

$$f'(x) = f(x)$$

"Oh, wow, okay. So we're looking for something that's equal to its own derivative?"

"That's right. This is an equality between a derivative $f'(x)$ and its original function $f(x)$. In other words, to solve this differential equation, we need to find a function $f(x)$ such that no matter what real number a you assign to x, $f'(a) = f(a)$ will be a true statement."

"And now I see what you mean about not being able to just integrate both sides, which in this case would result in something like this."

$$f(x) + C = \int f(x)\,dx$$

"If we don't know what $\int f(x)\,dx$ is, we don't know what $f(x)$ is!" Tetra said.

"We sure don't," I said.

"So... what can we do to solve this?" Tetra leaned forward to look more closely at the equations, making her sweet citrus scent all the more noticeable.

"Well, let's start with what we know. Think about the function $f(x) = e^x$, for example. We know that when we differentiate the exponential function e^x, its derivative is the same function, e^x. It doesn't change form."

$$(e^x)' = e^x$$

"In other words, the derivative of the function is the function itself."

$$f'(x) = f(x)$$

"So we can say that $f(x) = e^x$ is a particular solution to the differential equation $f'(x) = f(x)$."

Tetra wildly waved her right hand. "But hang on," she said. "I know about the exponential function e^x, and I know that $(e^x)'$ equals e^x. We used that when we talked about Taylor series, remember?"[1]

"We did, didn't we."

"But still, we're trying to solve the differential equation $f'(x) = f(x)$—trying to find a function $f(x)$, in other words—and out of the blue you just say 'how about e^x'? That kind of feels like cheating, like you already knew the answer."

"In a sense, sure, but for example when we were talking about $x^2 = 9$, you immediately told me that $x = 3$ was a solution, right?"

[1] See *Math Girls 2: Fermat's Last Theorem.*

"Well sure, because $3^2 = 9$."

"Same kind of thing here," I said. "You have to look at the form of the differential equation, and just see what you can come up with. Like, hey, an e^x would work here!"

Tetra crossed her arms. "I guess, but still..."

"By the way, can you think of any functions other than e^x that would satisfy $f'(x) = f(x)$?"

"Maybe something like $f(x) = e^x + 1$, since the e^x still won't change?"

"Yeah, but what about the 1?"

"Uh, let me write this out. So our function is this..."

$$f(x) = e^x + 1$$

"...and when we differentiate we get this..."

$$f'(x) = e^x$$

"So see? They're— Oh! Wait, that's *not* the same! So $f'(x) = f(x)$ doesn't hold in this case!"

"That's right. So $f(x) = e^x$ is a solution to this differential equation, but $f(x) = e^x + 1$ isn't."

"Well in that case I guess $f(x) = e^x$ is the only solution, since no matter what we add to it, the function and its derivative won't be the same."

"But how about something like $f(x) = 2e^x$?"

"Huh... I guess that works too, since differentiating $f(x) = 2e^x$ gives $f'(x) = 2e^x$. Which means something like $f(x) = 3e^x$ or $f(x) = 4e^x$ works too, right?"

"Of course. And again we can use a constant C to generalize solutions like that."

$$f(x) = Ce^x$$

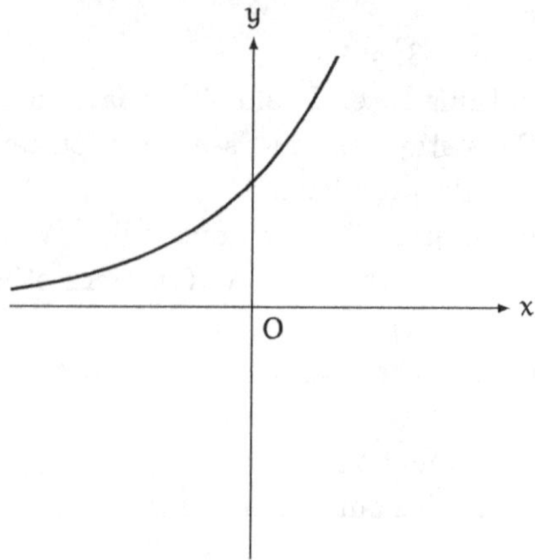

Graph of the particular solution $y = e^x$.

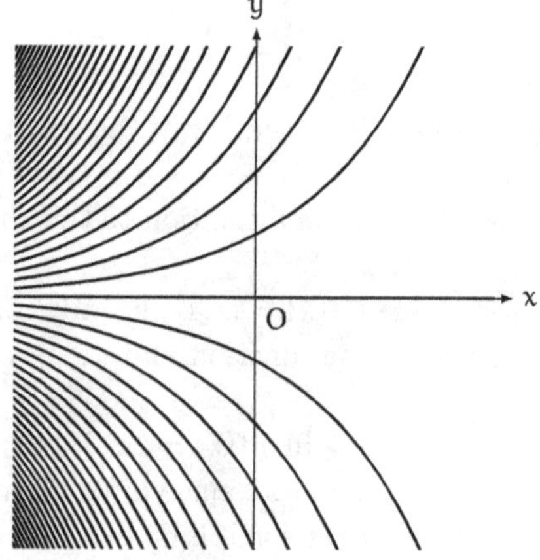

Sample graphs of the general solution $y = Ce^x$.

Tetra sat staring at my graphs, an unreadable expression on her face.

"What's wrong?" I asked. "You're following this, right?"

"Yeah, sure. But—sorry to repeat myself, but—I'm still hung up on what I was saying before."

"Namely?"

"I mean, I'm good with how integrating gave us the function $f(x) = 2x + C$. But it still feels like this particular solution $f(x) = e^x$

for the differential equation $f'(x) = f(x)$ came out of nowhere. That's the only way we got to the general solution $f(x) = Ce^x$, so... I don't know, it just makes me feel uneasy."

"Huh, okay. Then let's do this the right way." I paused to collect my thoughts, then resumed. "Right, so we want to solve this differential equation $f'(x) = f(x)$, which means our goal is to find this function $f(x)$. To make it clear what we're differentiating and why, let's let $y = f(x)$, making y a function of x. Then in our differential equation we can replace $f'(x)$ with $\frac{dy}{dx}$ and $f(x)$ with y, giving us this."

$$\frac{dy}{dx} = y$$

"So now our goal is to find y. Even if you don't like $y = e^x$ popping up 'out of the blue,' it's still important to consider what kind of function y is, right? You might say, for example, maybe y is a constant function, like this?"

$$y = C$$

"Well, you can find out if $y = C$ satisfies the differential equation $\frac{dy}{dx} = y$ by differentiating. Differentiating both sides of $y = C$ by x gives us this..."

$$\frac{dy}{dx} = 0$$

"...which tells us that the differential equation $\frac{dy}{dx} = y$ is satisfied only when $C = 0$. In other words, the constant function $y = 0$ is a particular solution to the differential equation $\frac{dy}{dx} = y$. Good so far, right? But now we need to consider what happens when $y \neq 0$.[2] Well in that case, since we know y can't be zero, we can divide both sides of the $\frac{dy}{dx} = y$ by y, giving us this."

$$\frac{1}{y} \cdot \frac{dy}{dx} = 1$$

"Now we can integrate both sides by x, giving us this."

$$\int \frac{1}{y} \cdot \frac{dy}{dx}\, dx = \int 1\, dx$$

[2]Non-constant functions having points such that $y = 0$ can be excluded by uniqueness of solutions to differential equations.

"Letting C_1 be our constant of integration, we have this."

$$\int \frac{1}{y} \cdot \frac{dy}{dx}\, dx = x + C_1$$

"On the left, we can use integration by substitution to get this."

$$\int \frac{1}{y}\, dy = x + C_1$$

"Assuming $y > 0$ and taking the indefinite integral of $\frac{1}{y}$, then with constant of integration C_2 we get this."

$$\ln y + C_2 = x + C_1$$

"Or, renaming $C_1 - C_2$ as $C_3 \ldots$"

$$\ln y = x + C_3$$

"From the definition of logarithms, we can say this."

$$y = e^{x+C_3}$$

"The right side becomes $e^{x+C_3} = e^x \cdot e^{C_3}$, and renaming e^{C_3} as C, we get this."

$$y = Ce^x$$

"We could also do all this with $y < 0$, and end up with the same thing. Also, letting $C = 0$, we get the particular solution of the constant function we started with."

$$y = 0$$

"And that's it. We've solved the differential equation $\frac{dy}{dx} = y$ with this general solution, same as before."[3]

$$y = Ce^x$$

"Okay," Tetra said. "You win."

[3]Strictly speaking, we also need to show that no solution with a different form exists (uniqueness of solutions to differential equations). For details, see Refs. [36] and [38] in the Recommended Reading section at the end of this book.

7.1.4 Trigonometric Functions

"As another example differential equation, how about this?" I said.

$$f''(x) = -f(x)$$

"This is the second derivative of $f(x)$, right?" Tetra asked. "Hmmm... I'm not sure. Using the exponential function we would have $(e^x)'' = e^x$, but we don't have the negative."

"One thing to notice is that since differentiating twice gave us a negative, four should bring us back to where we started. So, $f''''(x) = f(x)$."

"That just makes things harder!"

"You sure about that? I think you know a function that reverts to itself after differentiating four times..."

"Oh, wait, I do! The sine function!"

$$\begin{aligned}
(\sin x)' &= \cos x && \text{derivative of } \sin x \text{ is } \cos x \\
(\cos x)' &= -\sin x && \text{derivative of } \cos x \text{ is } -\sin x \\
(-\sin x)' &= -\cos x && \text{derivative of } -\sin x \text{ is } -\cos x \\
(-\cos x)' &= \sin x && \text{derivative of } -\cos x \text{ is } \sin x
\end{aligned}$$

Tetra cocked her head and looked at what I had written. "So the answer is $f(x) = \sin x$, right? We have $(\sin x)'' = (\cos x)' = -\sin x$, which means $f''(x) = -f(x)$, yeah?"

"Very good. So $f(x) = \sin x$ is one solution. Any others?"

"Others? Uh..."

"Well, for example we could use a constant A to create $f(x) = A\cos x$, or a constant B to create $f(x) = B\sin x$. So using A and B as parameters, we can represent the general solution like this."

$$f(x) = A\cos x + B\sin x$$

"Let me check this out!" Tetra said. "To confirm a solution to a differential equation, we differentiate both sides, right? So I should start with $f(x) = A\cos x + B\sin x$."

$$\begin{aligned} f'(x) &= (A\cos x + B\sin x)' \\ &= (A\cos x)' + (B\sin x)' \\ &= A(\cos x)' + B(\sin x)' \\ &= A(-\sin x) + B\cos x \\ &= -A\sin x + B\cos x \end{aligned}$$

"Okay, so now I differentiate $f'(x) = -A\sin x + B\cos x$."

$$\begin{aligned} f''(x) &= (-A\sin x + B\cos x)' \\ &= (-A\sin x)' + (B\cos x)' \\ &= -A(\sin x)' + B(\cos x)' \\ &= -A\cos x + B(-\sin x) \\ &= -A\cos x - B\sin x \\ &= -(A\cos x + B\sin x) \\ &= -f(x) \end{aligned}$$

"And I guess that does it! We've confirmed what we were looking for!"

$$f''(x) = -f(x)$$

I nodded. "Perfect."

7.1.5 The Why of Differential Equations

"Thanks to your examples, I think I'm starting to get a feel for what differential equations are," Tetra said. "Though I might still be a bit iffy on how to solve them."

Differential equation	General solution	
$f'(x) = 2$	$f(x) = 2x + C$	(for constant C)
$f'(x) = f(x)$	$f(x) = Ce^x$	(for constant C)
$f''(x) = -f(x)$	$f(x) = A\cos x + B\sin x$	(for constants A, B)

"To be honest, that's about all I know," I admitted.

"One thing, though..." Tetra said, her voice lower. "I still don't quite see the *why* of differential equations. What are they good for?"

There it is, Tetra's superpower. She may start with some pretty basic questions, but it isn't long before she really gets down to the essence of things. I'll bet that's how she learns so fast—she just naturally asks questions until her border between understanding and confusion becomes clear.

"An excellent question," I said. "I think the why of differential equations is the same as the why of 'normal' equations. After all, why do we set up an equation and solve it for x?"

"Because we want to know what x is, I guess?"

"That's right. Specifically, because we want to find any value or values for x that make the equation true. We know something about the properties of this x. Like, we might know it satisfies $x^2 = 9$. So we want to use those known properties to figure out what values x has."

"And you're saying differential equations are something similar?"

"Sure, the difference being we want to find a function instead of a value. We have this function $f(x)$, and we know something of its properties. That it satisfies the equation $f'(x) = 2$, for example, or $f'(x) = f(x)$ or $f''(x) = -f(x)$ or whatever. Regardless, the differential equation tells us something about the characteristics of this $f(x)$. We want to use that knowledge to help us figure out what the function could be. I think that's why we think about differential equations."

"Mmm..."

"That's a pretty powerful thing, finding a function," I said. "I mean, once you know a function, all you have to do is pass it an x to get a value from it. Then you can move the x around to see how $f(x)$ changes, you can investigate its asymptotic behavior when x gets really large, all kinds of things."

"I see."

I recalled what Ay-Ay had said about music being an art that plays out over time. "I guess finding a function can feel a little bit like receiving a prophecy."

"Prophecy?" Tetra said. "I find the idea of knowing the future a little bit scary. Like, something that maybe humans shouldn't be allowed to do."

"Well, there are limits to what we can know, of course. Errors, too."

"Still, prophetic functions? I don't know..."

"Okay, maybe 'prophecy' is taking things a little too far. It's not like we can learn *everything* about the future. But if we have a function of time for some physical quantity, we can know what that quantity will be in the future. The position of a star, for example. We can predict where in the sky that star will be thirty years from now. So, functions can be prophetic in that very limited sense."

At the time, thirty years in the future felt like an eternity. There were only a few months until my exams, and I was unable to predict anything about even such a short span of time.

"Functions representing physical quantities, huh. Any examples?" Tetra asked.

"Like... one describing the oscillation of a spring?"

7.1.6 Oscillations of Springs

"Say you have a spring, attached to the end of which is a weight with mass m, and you're stretching it out," I said. "At time $t = 0$, you carefully let go, and the weight begins to oscillate. If we can ignore friction, those oscillations will continue forever. Since the weight is oscillating, its position x will change with the passage of time. So the question is, exactly how will x change?"

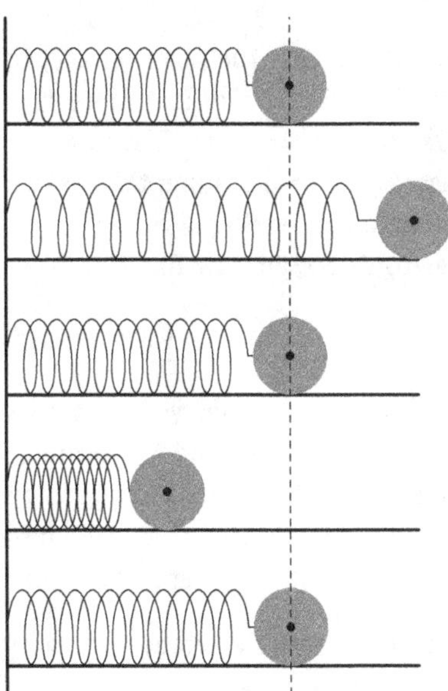

Oscillation of a spring.

"Actually, I've been referring to the position of the weight as x, but we can think of it as a function of time t, so we can write it as $x(t)$, right?"

"Sure, I see that," Tetra said.

"Great. So moving on, when solving problems in mechanics, we usually want to focus on forces. That's because once we understand the forces involved, we can apply Newton's second law of motion. Namely, if a force F is applied to a point mass m, then letting acceleration of that mass be α, we can write the relation $F = m\alpha$. This is a physical law."

$$F = m\alpha \qquad \text{Newton's second law of motion}$$

"I'm writing the force as F, but time t is hidden in there. Because as time passes, the force involved will change, right? And since F is changing with t, again, we need to think of the force as a function of time, $F(t)$. The acceleration α is a function of time, too. All good so far?"

Tetra nodded. "Go on."

"Okay, so if we differentiate the position $x(t)$ by time, we get the velocity $v(t) = x'(t)$, and if we differentiate the velocity by time we get the acceleration $\alpha(t) = v'(t)$. That means we can represent the acceleration as $x''(t)$. We'll assume the mass m is constant over time, so it won't change with t. So rewriting F as $F(t)$ and α as $x''(t)$, we can rewrite Newton's second law like this."

$$F(t) = mx''(t) \qquad \text{Newton's second law of motion (rewritten)}$$

"So, getting back to our spring, we want to think about what kind of force is acting between it and the point mass as the spring stretches and contracts. We can represent the force the spring exerts on the point mass as a function of this stretching and expanding. Here are some characteristics of the spring."

- When the spring is neither stretched nor compressed—in other words, when it's at its natural length—the force it exerts on the point mass is 0.

- When the spring is stretched from its natural length, there is a force acting in the direction opposite the direction of stretching and proportional to the amount of stretching.

- When the spring is compressed from its natural length, there is a force acting in the direction opposite the direction of compression and proportional to the amount of compression.

"These characteristics are a law of physics called Hooke's law. The weight's position determines the amount of stretching or compression of the spring, so we would naturally like to describe the relation between the force $F(t)$ the spring is applying to the mass and the mass's position $x(t)$. To simplify the representation, we'll let the weight's position when the spring is at its natural length be 0. Then we can represent Hooke's law like this."

$$F(t) = -Kx(t) \qquad \text{Hooke's law}$$

"What's this K?" Tetra asked.

"That's called the spring constant. It's a positive constant of proportionality that describes the stiffness of the spring. The bigger it is, the stronger the spring."

"And why the negative sign?"

"That shows the direction of the force, which operates in the direction opposite from the direction of stretching or compression."

"Ah, right. That was in the characteristics of the spring. Okay, I'm good."

"Great. So we've seen Newton's second law of motion and Hooke's law, both of which are related to the force $F(t)$ applied to the point mass."

$$\begin{cases} F(t) = mx''(t) & \text{Newton's second law of motion} \\ F(t) = -Kx(t) & \text{Hooke's law} \end{cases}$$

"Viewing this as a system of equations and getting rid of the $F(t)$, we get this."

$$mx''(t) = -Kx(t)$$

"Dividing both sides by m gives us this."

$$x''(t) = -\frac{K}{m}x(t)$$

"To make this easier to read, let's use an omega symbol to summarize the K and the m, rewriting $\frac{K}{m}$ as ω^2. We know that $\frac{K}{m} > 0$, so we can write this as a square. So anyway, that's about it. Now we have a differential equation for our function $x(t)$, which gives the position at time t of the weight attached to our spring."

$$x''(t) = -\omega^2 x(t) \qquad \text{Diff. eq. for spring oscillations}$$

"Take a close look at the form of this differential equation. Doesn't it look familiar?"

Tetra squinted at the page for a moment, then suddenly brightened. "Oh! It's just like this other one we did!"

$$f''(x) = -f(x)$$

"That's right. The only difference is the ω^2 coefficient, so we should be able to use the same trigonometric functions as before.

If we let $x(t) = \sin \omega t$, then we get $x''(t) = -\omega^2 \sin \omega t$, and the differential equation is satisfied. More generally, we can use the same constants A, B as before, and write it like this."

$$x(t) = A\cos \omega t + B \sin \omega t$$

"And there we have it—the general solution to our differential equation."

Tetra sat for a moment with a serious expression on her face, biting a fingernail as she silently reviewed everything I'd written. I expected her to raise a hand and ask a question at least once or twice, but instead she said, "Let me give this a shot." She pulled the notebook over to herself and started writing.

$$\begin{aligned} x(t) &= A\cos \omega t + B\sin \omega t && \text{general solution to the diff. eq.} \\ x'(t) &= (-A\sin \omega t)\cdot \omega + (B\cos \omega t)\cdot \omega && \text{differentiate both sides by } t \\ &= -\omega A\sin \omega t + \omega B\cos \omega t && \text{clean up} \end{aligned}$$

Tetra stopped there, thinking.

"Now you just have to differentiate one more time in the same way," I prompted.

"I know. Just thinking about something..."

She continued with her calculations.

$$\begin{aligned} x'(t) &= -\omega A\sin \omega t + \omega B\cos \omega t && \text{from above} \\ x''(t) &= -\omega(A\cos \omega t)\cdot \omega + \omega(-B\sin \omega t)\cdot \omega && \text{differentiate both sides by } t \\ &= -\omega^2 A\cos \omega t - \omega^2 B\sin \omega t && \text{clean up} \\ &= -\omega^2 (A\cos \omega t + B\sin \omega t) && \text{pull out the } -\omega^2 \\ &= -\omega^2 \underbrace{(A\cos \omega t + B\sin \omega t)}_{x(t)} && \text{found } x(t) \end{aligned}$$

"Sure enough, $x''(t) = -\omega^2 x(t)$." she said.

"Indeed, and you did a great job confirming that. What is it you were thinking about, by the way?"

"Oh, that. No big deal, just that you wrote the general solution using these constants A, B, and as usual the extra letters made me hesitate, thinking the problem had just become harder. But then I scolded myself, saying c'mon Tetra, they're just numbers. Nothing to be scared of here."

"Huh. That's a good way to look at it. Even when they're letters, they're still just numbers."

"One other thing, this $x(t)$, the position of the weight, that has a physical meaning, right? So I was thinking, what's the significance of A, B in terms of how the weight is moving?"

"And what did you come up with?"

"Well, looking at the general solution, when $t = 0$, that sort of determines A, B, doesn't it? Since we know that $\sin 0 = 0$ and $\cos 0 = 1$, we get this, right?"

$$\begin{aligned}
x(t) &= A\cos\omega t + B\sin\omega t && \text{equation for } x(t) \\
x(0) &= A\cos 0\omega + B\sin 0\omega && \text{substitute } t = 0 \\
&= A\cos 0 + B\sin 0 && \text{because } 0\omega = 0 \\
&= A && \text{because } \cos 0 = 1, \sin 0 = 0
\end{aligned}$$

"So it looks like A is determined as $A = x(0)$. And we can do something similar with $x'(t)$."

$$\begin{aligned}
x'(t) &= -\omega A\sin\omega t + \omega B\cos\omega t && \text{equation for } x'(t) \\
x'(0) &= -\omega A\sin 0\omega + \omega B\cos 0\omega && \text{substitute } t = 0 \\
&= -\omega A\sin 0 + \omega B\cos 0 && \text{because } 0\omega = 0 \\
&= \omega B && \text{because } \cos 0 = 1, \sin 0 = 0
\end{aligned}$$

"This time, B is determined as $\frac{x'(0)}{\omega}$!"

"Very nice!" I said.

"When I think of $x(t)$ as a function, I'm thinking in terms of math," she continued. "But when I think of $x(t)$ as telling me the position of the weight at time t, I'm thinking in terms of physics. So math is a kind of living language!"

"A living language?"

"Right. When we represent the position as $x(t)$, the function $x(t)$ has a physical meaning, allowing us to write Hooke's law as equations. But not just write it! We can move terms around and differentiate or whatever, but even so the equation retains its meaning, it keeps living on! Whatever we do, $x(0)$ still means the position of the weight at time 0, and $x'(0)$ still means the velocity at time 0. Setting $t = 0$ lets us observe what the weight is doing at time 0."

I nodded. "That's exactly right. The ω is determined by the mass m and the spring constant K, so the constants A, B are determined by the position and velocity of the weight at time 0. Then all you have to do is set a time, and you can know where the weight will be and how fast it will be moving at that time. This quantity we're calling ω also has a physical meaning, by the way. If you consider oscillations of the weight as a projection of circular motion, that motion will advance by the same angle over each fixed period of time, and ω tells us its angular velocity."

"That's so cool!" Tetra enthused, her hands clutched into fists in front of her chest. "I love how these equations have meaning, even when we change their form. A living language indeed, and a truly amazing one. It's like... like... creating meaning! I have *got* to write an article about all this for *Eulerians*."

"Your magazine, right."

"Differential equations almost feel like writing down parables that nature whispers to us. I can't wait to become better friends with them!"

Tetra was grinning, clearly excited by the prospect of becoming "friends" with yet another mathematical concept.

"You've made some pretty amazing progress over the past couple of years, you know."

"You bet! I can't keep letting a few extra letters hold me back!" She struck a heroic pose, pointing ahead. "Ever onward!"

I couldn't help but smile.

7.2 Newton's Law of Cooling

7.2.1 Afternoon Classes

My afternoon classes started. Well, not so much "classes" as just more study periods. Looking around, nearly everyone in the room was huddled over practice problems. I worked my way through two English comprehension exercises, then started daydreaming, thinking about what Tetra had said.

Explaining simple differential equations to her, using Newton's second law of motion and Hooke's law as examples, had reminded me of the close association between physics and mathematics. Both

$F = m\alpha$ and $F = -Kx$ are mathematical representations of physical laws, mathematics serving as the perfect "language" for accurately describing them. But mathematics isn't simply a language. By using its rules to change the form of equations, thereby deriving new ones, we can probe more deeply into their physical meaning. There is meaning not only in what we start with, but also in what we derive through transformations. Like Tetra said, mathematics is a living language.

I pulled out a physics text and started hunting for differential equations.

Problem 7-1 (Newton's law of cooling)

Given an object placed in a room with constant temperature U, let $u(t)$ be the object's temperature at time t. Assuming the rate of change of the object's temperature is proportional to the difference between its own temperature and the temperature of its surroundings (Newton's law of cooling), find $u(t)$ when the temperature at time $t = 0$ is $u_0 > U$, and the temperature at time $t = 1$ is u_1.

The most important thing here is to start with a mathematical description of the physical law in question. Since we're talking about rates of change, derivatives are the perfect tool.

- I'll use $u'(t)$ to represent the rate of temperature change.
- I'll use $u(t) - U$ to represent the temperature difference between the object and room temperature.

So I guess the most direct representation of Newton's law of cooling, that the rate of temperature change between an object and its surroundings is proportional, would be this.

$$u'(t) = K\big(u(t) - U\big) \qquad \text{for constant } K$$

So that's the world of physics. Now for the world of mathematics. But to get there, I need to build a bridge.

I want to know the function $u(t)$. Specifically, I know U, u_0, and u_1, so I want to represent $u(t)$ using those building blocks.

I could transform this equation into a differential equation I knew.

$$u'(t) = K\big(u(t) - U\big) \qquad \text{Newton's law of cooling}$$
$$\big(u(t) - U\big)' = K\big(u(t) - U\big) \qquad \text{rewriting } u'(t) \text{ on the left}$$

I rewrote the left side like this to get the same $u(t) - U$ on both sides.

$$\big(\underset{\sim\sim\sim}{u(t) - U}\big)' = K\big(\underset{\sim\sim\sim}{u(t) - U}\big)$$

By doing so, I could create a differential equation in exactly the same form as one I'd talked about with Tetra.

$$f'(t) = f(t)$$

The general solution should involve the exponential function, but when I differentiate I need this coefficient K to show up. I need to get not Ce^t but Ce^{Kt}.

$$\underset{\sim\sim\sim}{u(t) - U} = Ce^{Kt} \qquad \text{for constants } C, K$$

To confirm, I'll differentiate both sides with respect to t.

$$\begin{aligned}\big(u(t) - U\big)' &= Ce^{Kt} \cdot K \\ &= KCe^{Kt} \\ &= K\big(u(t) - U\big)\end{aligned}$$

Sure enough, this satisfies the differential equation. So now I had $u(t) - U = Ce^{Kt}$, in other words...

$$u(t) = Ce^{Kt} + U \qquad \cdots (1)$$

Now I know a good bit about this function $u(t)$ describing the object's temperature at time t. What I still don't know, however, are these two constants, C and K.

I took a moment to reread the problem, and noticed two conditions I'd been forgetting.

- The temperature at time $t = 0$ is u_0.
- The temperature at time $t = 1$ is u_1.

So to find C and K, I need to look at this equation (1) when $t = 0$ and when $t = 1$.

$$\begin{cases} u_0 = Ce^{K \cdot 0} + U & \cdots (2) \\ u_1 = Ce^{K \cdot 1} + U & \cdots (3) \end{cases}$$

Finding C is straightforward. Since $e^{K \cdot 0} = e^0 = 1$ in equation (2), I get $u_0 = C + U$. In other words, C is this.

$$C = u_0 - U$$

Substituting this C into equation (3), I get this.

$$u_1 = \underbrace{(u_0 - U)}_{C} e^{K \cdot 1} + U$$

Now with a bit more calculation, I can find e^K.

$$\begin{aligned} u_1 &= (u_0 - U)e^{K \cdot 1} + U \\ u_1 - U &= (u_0 - U)e^K \\ e^K &= \frac{u_1 - U}{u_0 - U} \end{aligned}$$

The problem states that $u_0 > U$, in other words $u_0 - U \neq 0$, so I know it's safe to divide by $u_0 - U$.

So now I've found e^K. All that's left is summarizing everything as equations.

$$\begin{aligned} u(t) &= Ce^{Kt} + U \\ &= (u_0 - U)e^{Kt} + U \\ &= (u_0 - U)(e^K)^t + U \\ &= (u_0 - U)\left(\frac{u_1 - U}{u_0 - U}\right)^t + U \end{aligned}$$

So here I've used u_0, u_1, and U to represent $u(t)$.

$$u(t) = (u_0 - U)\left(\frac{u_1 - U}{u_0 - U}\right)^t + U$$

Now some test calculations, to make sure $u(0) = u_0$.

$$\begin{aligned} u(0) &= (u_0 - U)\left(\frac{u_1 - U}{u_0 - U}\right)^0 + U \\ &= (u_0 - U) + U \\ &= u_0 \end{aligned}$$

There it is. Next, does $u(1) = u_1$?

$$\begin{aligned} u(1) &= (u_0 - U)\left(\frac{u_1 - U}{u_0 - U}\right)^1 + U \\ &= (u_0 - U)\left(\frac{u_1 - U}{u_0 - U}\right) + U \\ &= u_1 - U + U \\ &= u_1 \end{aligned}$$

Yep.

Answer 7-1 (Newton's law of cooling)

Given an object placed in a room with constant temperature U, let $u(t)$ be the object's temperature at time t. Assuming the rate of change of the object's temperature is proportional to the difference between its own temperature and the temperature of its surroundings (Newton's law of cooling), $u(t)$ is

$$u(t) = (u_0 - U)\left(\frac{u_1 - U}{u_0 - U}\right)^t + U,$$

where $u_0 > U$ is the temperature at time $t = 0$, and u_1 is the temperature at time $t = 1$.

So there's my answer. Time to head back to the world of physics. What can I learn from the form of this temperature function $u(t)$?

I'd found this bit $u_0 - U$. I know u_0 is the temperature at time 0, and U is the room temperature, so $u_0 - U$ is the starting difference in temperature between the object and its environment. The conditions state that $u_0 > U$, so $u_0 - U > 0$.

So how should I read this?

$$\left(\frac{u_1 - U}{u_0 - U}\right)^t$$

Overall, this is an exponential function of time t. In any case, this is the part equivalent to the e^{Kt} I was after.

Just because it's an exponential function, that doesn't necessarily mean it will increase. Since $u_0 > U$, as the time t moves from 0 to 1, the object's temperature $u(t)$ should be approaching the room temperature U. It couldn't go beyond room temperature, however, so $u_1 > U$.

$u_0 - U$ and $u_1 - U$ are both positive. Further,

$$u_0 - U > u_1 - U > 0.$$

That means I could say this.

$$0 < \frac{u_1 - U}{u_0 - U} < 1$$

In other words,

$$\left(\frac{u_1 - U}{u_0 - U}\right)^t$$

should get closer to 0 as t increases.

I need to graph this as $y = u(t)$.

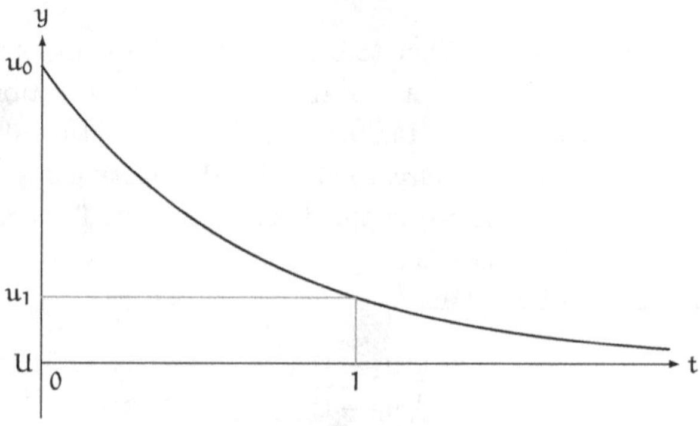

A graph of $y = u(t)$.

This shows how the temperature of the object approaches room temperature. I had used the laws of physics to set up a differential equation, solved it, and thereby learned how a physical quantity changes over time.

A living language indeed.

I flipped through the textbook some more, and came across a section on radioactive decay.

Problem 7-2 (radioactive decay)

Let $r(t)$ be the residual amount of a radioactive material at time t. Find $r(t)$ in terms of the residual amount r_0 at time $t = 0$ and the residual amount r_1 at time $t = 1$, assuming the rate of radioactive decay is proportional to the residual amount.

Huh, this is the same thing. The same as Newton's law of cooling.

Okay then, let's mathematically represent radioactive decay. Again, there's a rate of change, so I can use derivatives.

- I'll use $r'(t)$ to represent the rate of decay of a radioactive material.
- I'll use $r(t)$ to represent the residual amount of the radioactive material.

So I can represent the rate of radioactive decay being proportional to the residual amount like this.

$$r'(t) = Kr(t) \qquad \text{for constant } K$$

An object cooling down and radioactive decay are two completely different physical phenomena. The physical quantities their functions represent are temperature and mass, which again are completely different. Even so, the form of the differential equations describing these physical quantities is the same. This of course implies the functions that are their solutions will also have the same form.

So I can just replace $u(t) - U$ with $r(t)$, $u_0 - U$ with r_0, and $u_1 - U$ with r_1.

$$u(t) - U = (u_0 - U)\left(\frac{u_1 - U}{u_0 - U}\right)^t \qquad \text{Newton's law of cooling}$$

$$r(t) = r_0 \left(\frac{r_1}{r_0}\right)^t \qquad \text{radioactive decay}$$

Letting $U = 0$ here gives exactly the same form.

$$u(t) = u_0 \left(\frac{u_1}{u_0}\right)^t \qquad \text{Newton's law of cooling (with } U = 0)$$

$$r(t) = r_0 \left(\frac{r_1}{r_0}\right)^t \qquad \text{radioactive decay}$$

So the "living language" of mathematics is revealing how two very different phenomena actually have a common behavior. In this case, it is the forms of differential equations and their solutions that provide that revelation.

Answer 7-2 (radioactive decay)

Assuming the rate of decay of a radioactive material is proportional to its residual amount, the residual amount at time t is

$$r(t) = r_0 \left(\frac{r_1}{r_0}\right)^t,$$

where r_0 is the residual amount at time $t = 0$, and r_1 is the residual amount at time $t = 1$.

Tetra had described differential equations as parables that nature whispers to us. *Such an interesting perspective... Hang on, this radioactive decay problem is describing the half-life of radioisotopes. Which means there's also something like a half-life for temperature change... Huh...*

I passed the rest of my study period with my mind wandering among similar paths and byways. And so flowed by yet another day.

The rate of decay of a radioactive material is proportional to its residual amount.

Rate of radioactive decay

CHAPTER 8

A Remarkable Theorem

> Long before the time of Gauss, it was suggested by J.H. Lambert (1728–1777) that, if a non-Euclidean plane exists, it should resemble a sphere of radius i.
>
> H.S.M. COXETER
> *Introduction to Geometry [21]*

8.1 AT THE STATION

8.1.1 Yuri

Leaving the train station on my way home from school one evening, I bumped into Yuri. I only rarely ran into her on my way to school, and seeing her on my way home was rarer still.

"Fancy meeting you here," I said.

"Yo, cuz! Heading my way?" she said.

We crossed the pedestrian bridge in front of the station. We lived in practically the same neighborhood, so our routes home were nearly identical.

"Have you grown?" I asked.

"Larger and larger, every time you see me."

"Practically a giant, you are," I said, giving her a playful bump on the back with my bag.

"Ouch!"

We turned off the main street, leaving the traffic noise behind us as we entered a quiet residential area.

"Oh, I have a new problem for you!" Yuri said.

Yuri's problem

- You start at point A and walk some distance in a straight line to point B. You turn $\frac{\pi}{2}$ to the left.
- From point B you walk the same distance in a straight line to point C, and again turn $\frac{\pi}{2}$ to the left.
- From point C you walk the same distance in a straight line and arrive back at point A, and again turn $\frac{\pi}{2}$ to the left.
- You find yourself facing the same direction as when you started.

Find the area of the triangle your walk described.

"Hang on a second," I said. "There's something strange about this."

"If it's the $\frac{\pi}{2}$ that's bothering you, that just means 90°. We're using radians, not degrees here. 180° is π radians, and 360° is 2π radians. Figured you'd know that, math boy."

I sighed. "Not what I meant. You said I'm visiting three points A, B, C, and each time I'm turning by $\frac{\pi}{2}$, or 90°. So how is that a triangle?"

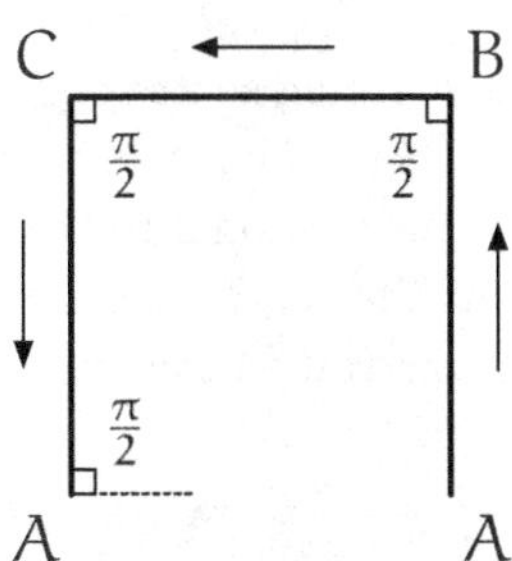

"Giving up already?" Yuri said with an I-win grin.

Ah, okay.

"Not so fast," I said. "I see what's going on. I'm walking on a sphere, not on a plane, right? That's why there can be a triangle with three $\frac{\pi}{2}$ angles."

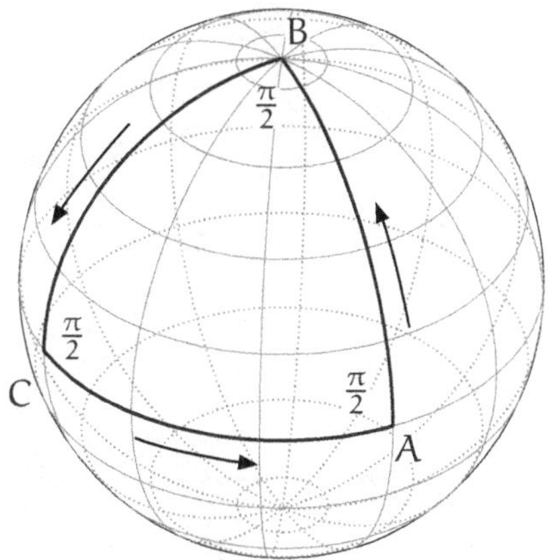

Yuri clucked. "Got it already, huh. I should have known."

"Right, if we consider a 'straight line' on a globe as a path following a geodesic, a great circle in other words, it's possible to create a triangle with all right angles. In spherical geometry, the size of a triangle determines the size of its angles. This is a pretty cruel problem, though. You're making me walk the distance from the equator to the north pole for each side of the triangle. I'm exhausted."

"I didn't say anything about you walking on the earth."

"Then your problem is lacking some conditions. You haven't told me the radius R of the sphere. That's important, because the area of the triangle should be proportional to the square of R."

"We'll say figuring out the conditions is part of the problem."

"That doesn't sound like you," I said. "You're normally all about the details."

Yuri's problem, revised (area of a spherical triangle)

Find the area Δ of a triangle △ABC on a sphere with radius R and having angles

$$\angle A = \angle B = \angle C = \frac{\pi}{2}.$$

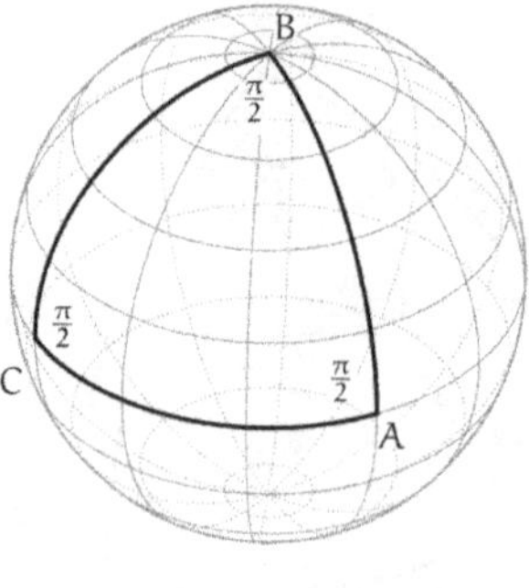

"So do you have an answer yet?" Yuri asked.

"Easy. Eight of these triangles would exactly cover the entire sphere. On a globe, you'd have four in the northern hemisphere and four in the southern. The surface area of a sphere with radius R is $4\pi R^2$, so just multiply that by $\frac{1}{8}$, giving $\frac{\pi R^2}{2}$."

Answer to Yuri's revised problem (area of a spherical triangle)

The surface area of a sphere with radius R is $4\pi R^2$, and the area of $\triangle ABC$ is equivalent to $\frac{1}{8}$ the surface area of the sphere. The area Δ of the triangle in question is therefore

$$\Delta = \frac{4\pi R^2}{8} = \frac{\pi R^2}{2}.$$

8.1.2 Surprise

"Okay, I guess that one was too easy for you," Yuri said. "By the way, you can also calculate the area like this."

$$\Delta = R^2\left(\frac{\pi}{2} + \frac{\pi}{2} + \frac{\pi}{2} - \pi\right)$$

"Or so I hear, at least."

I was a bit confused, unsure where this came from.

"What's this all about?" I asked.

"Just like I said. You can add the three angles together, subtract π, and multiply all that by R^2."

$$\begin{aligned}\Delta &= R^2(\angle A + \angle B + \angle C - \pi)\\ &= R^2\left(\frac{\pi}{2} + \frac{\pi}{2} + \frac{\pi}{2} - \pi\right)\\ &= \frac{\pi R^2}{2}\end{aligned}$$

"And where does this come from?"

"Well you're the one who told me congruence and similarity are the same in spherical geometry, right? So I used that in a problem for that guy, and he's like 'well of course' and all. Because determining the sizes of the angles determines the form of a spherical triangle, and that determines its area. So that's the formula for determining the area from the three angles! That's what he said, at least."

Area of a spherical triangle

The area Δ of a spherical triangle $\triangle ABC$ on a sphere with radius R is

$$\Delta = R^2 (\angle A + \angle B + \angle C - \pi).$$

"I see," I said, perplexed. "He" would be Yuri's boyfriend. He had transferred to a different junior high school, but apparently they still met from time to time. Apparently they were engaged in some unremitting math battle, but I wasn't sure of the details. More importantly, however, I wasn't sure how such a simple formula could really give the area of a spherical triangle.

"Oh, right! I just remembered another surprising thing, that— Oh, oops. We're here already."

I looked up and realized that while my mind was off in mathland we'd arrived in front of my house.

"What's the surprise?" I asked.

"We'll pick it up again next time, cowboy." Yuri closed one eye and made her hands into guns. "Pew-pew!"

I entered my house, my head filled with a too-simple formula and my body riddled with imaginary bullets.

8.2 At Home

8.2.1 *Mother*

Yuri's formula for the area of a spherical triangle was still bouncing around in my brain as I helped set the table for dinner. According to her, letting angles $\angle A, \angle B, \angle C$ respectively be α, β, γ, the area Δ would be

$$\Delta = R^2 (\alpha + \beta + \gamma - \pi).$$

It just seemed too simple. If it really worked, there had to be something interesting going on there.

Situations like this were my favorite part of doing math. Not being given a problem and finding the right answer. Not being given a theorem and seeing how to prove it. It was this search for "something interesting" that I loved. Some new thing I could derive using logic, something with mathematical significance.

"I hope Miruka drops by again soon," my mother said, bringing a large salad bowl to the table. "I want to hear more about triangles."

"You'd have to ask her," I said. "She was really singing your praises the other day, by the way."

"You don't say? Tell me more!"

I told her more, but my head was still filled with math. I wanted to quickly finish my dinner and put my thoughts down on paper. Until then, I just made a series of mental notes.

I could see how the area of a spherical triangle would be proportional to R^2. What struck me as odd was the $\alpha+\beta+\gamma-\pi$ part—it didn't seem like angles could so directly lead to areas—so I decided to focus on that.

For example, what happens if I divide by R^2*?*

$$\Delta = R^2(\alpha+\beta+\gamma-\pi)$$

$$\alpha+\beta+\gamma-\pi = \frac{1}{R^2}\Delta \qquad \text{exchange and divide by } R^2$$

"Whoa!" I shouted.

My mother shouted too, in surprise. "What's wrong? Too hot?"

I realized I was sitting across the table from my mother, and that I'd just swallowed a mouthful of pumpkin soup. I'd been so preoccupied by math that I'd been eating dinner like an automaton.

"Sorry. No, I'm fine. Just made something of a discovery."

"You startled me! So anyway, as I was saying—"

I kept thinking as my mother's words went in one ear and out the other.

So what happens if I keep the area of $\triangle ABC$ *fixed while* $R \to \infty$*?*

$$\alpha+\beta+\gamma-\pi = \frac{1}{R^2}\Delta$$

The $\frac{1}{R^2}$ here meant that the limit as $R \to \infty$ should be this.

$$\alpha+\beta+\gamma-\pi = 0$$

In other words, this.

$$\alpha+\beta+\gamma = \pi$$

π radians is 180°, so... So this is just taking me back to elementary school geometry, saying that the sum of the interior angles of a triangle is 180°!

Okay, now things were starting to make sense. Taking the limit as $R \to \infty$ means considering larger and larger spheres. I imagined a truly enormous inflating sphere, and could see how its surface would become increasingly like a plane. So it felt natural to consider the properties of a triangle drawn on a sphere to start approximating those on a plane as the sphere's radius approached infinity.

The formula for area of a spherical triangle

$$\Delta = R^2 (\alpha + \beta + \gamma - \pi)$$

that had seemed so strange at first now felt much more familiar, like a stranger becoming a friend. Turns out it was just the spherical version of what I already knew about planar triangles! Or so I assumed. From somewhere, a voice was whispering to me—

Prove it.

Problem 8-2 (area of a spherical triangle)

Show that the area Δ of a triangle $\triangle ABC$ on a sphere with radius R is

$$\Delta = R^2 (\alpha + \beta + \gamma - \pi),$$

where α, β, γ are the interior angles of $\triangle ABC$.

"—don't you think? Hey, are you even listening to me?" my mother said.

"Um, sorta." I shoveled down the rest of my dinner and hurried to my room.

8.2.2 Rare Things

As it turned out, finding areas of spherical triangles wasn't as simple as I'd hoped.

Things were easy in special cases, like Yuri's problem where $\alpha = \beta = \gamma = \frac{\pi}{2}$, but I found myself flummoxed when trying to find the area of *any* spherical triangle. Given a sphere floating in three-dimensional space, using equations in x, y, z to describe its great circles and calculate angles, how would I find areas? Using integrals? That sounded... hard.

I looked at the calendar I'd placed on the front edge of my desk, marking my schedule leading up to entrance exams. Next to that was a schedule by subject.

So little time.

Yuri's problem had revealed a trail I very much wanted to explore. Extensively. But I didn't have the time.

I sighed and put away the papers I'd been scribbling on, pulling out the results from my mock exam I'd gotten back the other day—tables showing points awarded, standardized scores, and rankings by subject. I'd gotten a "B" ranking for entrance to the university I most wanted to attend. Partial credit on the math section, sure enough. If only I hadn't forgotten that substitution, I would have gotten a perfect score. *And I wouldn't have embarrassed myself in front of Miruka...*

Particularly bad was my score on the classical literature section, where I'd lost far more points than I'd expected on the comprehension questions. Definitely a weak point I needed to shore up. My average score across all subjects wasn't bad, but I couldn't let one weak subject hold me back from the "A" ranking I wanted.

As I reviewed the comprehension problems I'd missed, I wondered why someone like me, someone who knew they were heading for a math or science major, needed to read classic literature. To be tested on it, even!

But things were what they were. I flipped to one of the problems I'd missed. The 75th stanza of *The Pillow Book of Sei Shonagon*, an excerpt from one of her "listing poems."

> One demonstrating excellence of appearance, character, and attitude, always dealing with the world in a flawless manner.

I had no idea what she wanted to say, but then I looked at the header above it and saw that this was from a poem titled "Rare Things." So I guess she wanted to say almost nobody is like this? An old

Japanese proverb says "the heavens are stingy in their gifts," but they were certainly generous in some cases.

I was of course thinking of the brilliant and beautiful Miruka. Not the superficial stuff like how she got excellent grades and had a pretty face, but how she was so deep, and so strong. Compared to her, I was nothing. There's a time parameter t hidden in "dealing with the world." I imagined moving that t forward, how doing so would only increase the differences between us.

I sighed again. *Thoughts like this aren't helpful. All I can do is all I can do.*

I would be taking my final mock exam soon. I needed to work on the areas where I needed improvement and get that "A" ranking. Perhaps the heavens had given me only a few gifts, but I could continue with my studies, I could take my mock exams, I could get a feel for how well I'm doing. I could do all I could do to get ready.

8.3 In the Library

8.3.1 Tetra

The skies were remarkably clear the next day, but it was very cold.

After classes I headed to the library as usual, intending to review some classical literature vocabulary and practice a few comprehension exercises. Passing through the door, I almost collided with the red-haired Lisa.

"Sorry," she muttered, clutching her crimson notebook computer to her chest as she scurried past. Looking into the library, I saw Tetra sitting at a table near the window. She had an uncharacteristically dejected look on her face. I walked up to her table and took a seat. Something about her smelled different...

"You okay?" I said. "I just passed Lisa heading out, practically running..."

"Yeah," Tetra said. "We had... I don't know. A difference of opinion."

"A difference of opinion?"

"About *Eulerians*, yeah. I had a bunch of ideas for articles, but she just shot down every one. She didn't even consider them!"

"Ah, I see. A matter of editorial policy, then."

"I guess. But there are so many things I want to write about!" she said. She started ticking off subjects on her fingers. "Continuity and topological spaces and epsilon–delta and open sets and homeomorphisms and spherical geometry and Euclidean geometry and hyperbolic geometry and... And I wanted to present it all with examples and equations and figures, along with related concepts and historical connections and how they're related to other areas in mathematics and... Well anyway, I wanted to write articles about all that stuff."

"The bridges of Königsberg, maybe?"

"Oh, absolutely!" She started flipping through a thick notebook. "I can't write about topology and skip that!"

There was something about this scene that felt familiar.

"Uh, Tetra," I said. "Maybe you're trying to put in a little too much? Remember your presentation on randomized algorithms, how you had to cut way back on that?"

"Well yeah, but you need it all! Everything builds up! You can't just start in the middle, the readers wouldn't understand a thing! And geometry is so *interesting.* There are so many things to talk about regarding shapes—magnitude, congruence, similarity, straight lines, curves, angles, areas..."

"Yeah, but—"

"Everything I've learned about topology and non-Euclidean geometry has been so surprising, how concepts I've always thought to be unchangeable actually aren't. Concepts I've never even *considered* changing. Straight lines that aren't really straight, and distances without length, and—"

"Sure, I can see that, but—"

"I want to write all this stuff down in a way that captures how much *fun* it all is. But you've got to do the job right to get all that across, right? But when I tell Lisa all that, she just brushes it all aside." Tetra's frown deepened.

"So..." I began, as softly as possible. "I'm not sure how long you were intending to make *Eulerians*, but sometimes writing down *everything* can actually make it harder to get your point across."

"What do you mean?"

"Well, you're a great reader and a great writer, and you're also really good at both math and languages, but not everyone is like that. If you try to dump everything on them at the same time, your readers are more likely to get lost. There's no way to get your message across if that happens, right?"

"Well yeah, but still..."

"You said you want to let people know how fun all this stuff is. But you don't have to pour out all the fun at once. You need to be a little more selective. You said something interesting the other day, you know."

"What was that?"

"You said creating differential equations is like writing down parables that nature whispers to us. You also called mathematics a living language. Both of those really stuck with me."

"I had no idea."

"You're really good at putting your understanding into words. So I'd say, rather than trying to express *all* of your understanding, maybe you should focus on just one topic. Then if you do a good job of writing at your border between understanding and confusion, you'll end up with a very interesting *Eulerians*."

"I guess, maybe..."

"I bet Lisa is thinking something similar, and that's why she's rejecting your ideas."

"Hard to say. All she ever says is 'no' when I ask her what she thinks."

"Well, Lisa isn't exactly the most skilled communicator. Maybe you just need to take some time with her, figure out what it is she's trying to say."

"Take some time?"

"Remember when you were talking about how through cooperation, you can accomplish things you can't do alone? There are things you're good at, and there are things Lisa's good at. Cooperating means bringing out each other's strengths and complementing for each other's weaknesses."

"I see. So you're saying I should try to make up for what Lisa isn't so good at?"

"Sure, and that she can do the same for you. When you can both do that, I think you'll create an *Eulerians* neither of you could have made on your own."

8.3.2 Stating the Obvious

Tetra stared off into space, thinking about what I had said.

"Sorry if I'm coming off as pretentious, or just stating the obvious," I said.

"Not at all. Like you say, it's best to start from the obvious," Tetra said, her frown now replaced by a serious expression. "And you're right, I probably should have tried harder to figure out what Lisa was trying to say."

"Speaking of stating the obvious, I just learned something surprising about triangles, of all things..."

Recalling my excitement from the previous night, I told Tetra about the spherical triangle problem I'd been working on. When I was done, Tetra's eyes were shining.

"You can find the area of a spherical triangle from *this*?"

$$\Delta = R^2 (\alpha + \beta + \gamma - \pi)$$

"Apparently so," I said. "Also interesting is that a spherical triangle approximates a planar triangle when you take the limit as $R \to \infty$. Which I guess isn't so surprising, since making the radius larger makes the sphere more like a plane."

"So the Flat Earthers would be right, if Earth was infinitely large!" Tetra laughed. "Oh, not to change the subject, but check this out! The infinity sign!"

Tetra stood from her seat. She touched her right thumb to her left index finger and vice versa, awkwardly holding them out for me to see. Sure enough, if you looked at it the right way the loops formed by her crossed fingers looked something like an ∞ symbol.

"This is a new hand sign, I guess?"

"Right! You make the sign in front of your chest, then push it out toward whoever you're sending the sign to and announce, 'Infinity!' Like this." She directed one toward me. "Infinity! ... What do you think?"

"Sure, that's great," I said. (Honestly, I thought it was adorable.)

"Infinity!" Tetra repeated, clearly enjoying herself.

"But back to triangles, if I may," I said. "In the end I wasn't able to actually prove the area would be $R^2 (\alpha + \beta + \gamma - \pi)$. I'm pretty sure there are integrals involved, so it looked like a pretty tough nut to crack. So I set it aside—" I almost said "until I have more time," but held back. "—for now."

"I'll bet Miruka knows," Tetra said. *Ouch.*

"I guess. Haven't seen her around school, though. I guess she's back in the States again."

Back in the States, where I can't apologize for my behavior. Where the timing for doing so can only get worse. I wondered if things between us would remain the same throughout my exams. And graduation...

"Miruka?" Tetra said. "She's around. She just dropped by and and gave me some chocolate. See?"

Tetra held up a small bag of candies. *So* that's *what I smelled...*

"Uh, when was this?"

"About thirty minutes ago, I guess? She said she was heading to the student lounge. These are super yummy. Want one?"

"Sorry, there's something I have to do."

I rushed for the door.

8.4 At the Student Lounge

8.4.1 Miruka

I raced down the path leading to the building where our student lounge was. There were several small groups of students there, some having whispered meetings about club business, others studying on their own.

Miruka was off alone in a corner, reading a book. The table she was sitting at was covered in papers, one of which she was writing on with a fountain pen. Just seeing her, I felt speechless. It was almost as if there was a barrier around her, making her unapproachable.

I'm such a fool. How could I have thought Miruka's brilliance was a "gift from the heavens"? Obviously, she knew so much because she studied hard.

The Miruka who gave us lectures wasn't all of Miruka. Just as she had said, what I saw of her was not all of her. She also spent time reading, and thinking, and working on problems, and that's how she'd become who she was. All that time I'd spent whining, she'd been learning. While I'd been— *Oops, she's looking this way...*

She gave a slight smile and crooked an index finger. I headed toward where she was sitting, as if pulled by an invisible thread.

8.4.2 Words

"There you are," Miruka said. "*Meinen Brotkrümeln folgen.*"

I had no idea what that meant.

"What are you reading?" I asked, the lamest thing I could possibly say.

"A book Dr. Narabikura recommended." She tilted it up so I could see the cover. The title was in English, indecipherable to me.

"Something to help you decide on a field of study?"

"No, I'm nowhere near that yet. I'm still covering the basics before I move on to the really hard stuff, making sure I'm solid on the English first. This is a textbook I'll be using in college next year, so I'm getting a little sneak peek for practice." She shrugged. "That, and the occasional paper that catches my eye."

"Nice," I said, wondering if I could make my way through the alternate universe of a math text written in English.

"There's a seminar starting up soon, so I'm also supposed to be on the lookout for interesting topics as I learn. If anything promising pops up, I'll try to write a paper about it. Then more papers, and ever more. Euler wrote countless papers. I'm hoping to give him a run for his money. Papers are how I'll leave behind what I've thought, what I've discovered. I'll be writing them for myself, and as something to pass on to the next generation. Also to pass along what Dr. Narabikura has given me."

"Lots of 'done and done,' I guess."

Miruka smiled at me. "That's the plan. Papers are letters to the future. I hope to be a very prolific penpal."

Looking back and forth between her and the blue-book notebook she'd been filling up with equations, I felt my chest tighten. Miruka wasn't faced with college entrance exams like I was, but she had plenty on her plate even so, her own preparations for entering a new world.

"You're amazing," I said.

"Where'd that come from?"

"Seriously. You're about to fly off to a world I'll never reach. I'm just glad I got to meet you before you do."

Miruka looked away from me, toward a window. I'd just told Tetra to listen more closely to what Lisa was trying to tell her. But what about me? Was I really listening to Miruka, one of the most important people in my life? She had a good grasp on the future she wanted, and was heading straight toward that future without over-estimating or under-estimating her abilities.

Was I really, truly listening to what she was telling me?

8.4.3 Solving Riddles

"Enough about me," Miruka said, returning her gaze. "What are you up to?"

"Ah, right. What do you know about spherical triangles? I've been working on a proof, but I'm kinda stuck."

Problem 8-2 (area of a spherical triangle)

Show that the area Δ of a spherical triangle $\triangle ABC$ on a sphere with radius R is

$$\Delta = R^2(\alpha + \beta + \gamma - \pi),$$

where α, β, γ are the interior angles of $\triangle ABC$.

"Hmph. Well, let's start from the obvious," Miruka said. She started sketching a ball. "We know the surface area of a sphere with radius R is $4\pi R^2$. So if you draw a single great circle on a sphere, you create two hemispheres, each with a surface area of $2\pi R^2$. Draw two different great circles, and you divide the sphere into four wedge shapes. You can consider these as '2-gons' in spherical geometry, but we'll call them 'lunes.' Each lune has two equal angles, so we'll refer to a lune with angles of magnitude α as an α-lune. When you use two great circles to create four lunes, you'll have two α-lunes and two $(\pi - \alpha)$-lunes. So our first question is, what's the surface area of a single α-lune?"

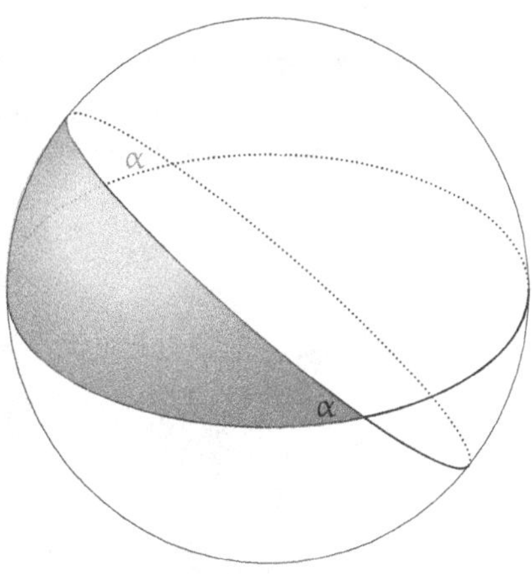

An α-lune.

"Letting S_α be the surface area of one α-lune, S_α will be proportional to α. A π-lune is a hemisphere, so its surface area is this."

$$S_\pi = 2\pi R^2$$

"That means the constant of proportionality is $2R^2$, giving us this."

$$S_\alpha = 2\alpha R^2$$

"So now we've described the surface area of an α-lune in terms of R and α. Good with all that?"

I nodded. "Yeah, sure."

"Okay. So if you draw *three* different great circles on a sphere, you'll divide it into spherical triangles. When you do so, what's the relation between a spherical triangle $\triangle ABC$ and the α-, β-, and γ-lunes?"

Miruka stopped there, clearly passing the baton. Physically as well as metaphorically, since she was holding out her fountain pen for me to use.

Taking the pen, I looked at her figure and started thinking on how to proceed. I knew the surface area of the sphere, and I knew the surface area of a lune, so how can I build on that? I stared at her sketch a little longer, but nothing came to me.

"You aren't going to get anywhere by just staring," Miruka said. "Gotta get those fingers moving." She snatched her pen from my hand and started working on some new graphs.

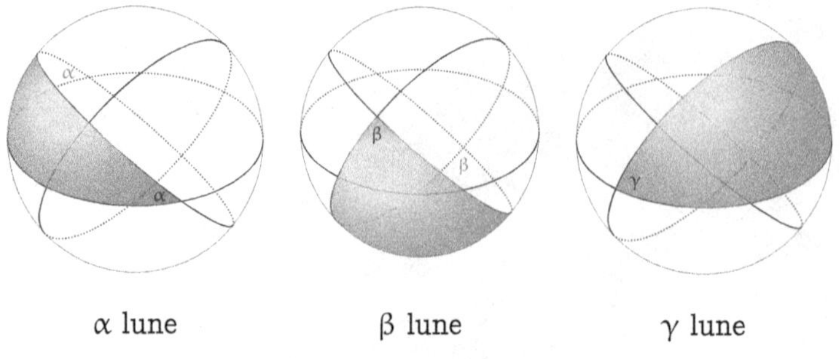

α lune β lune γ lune

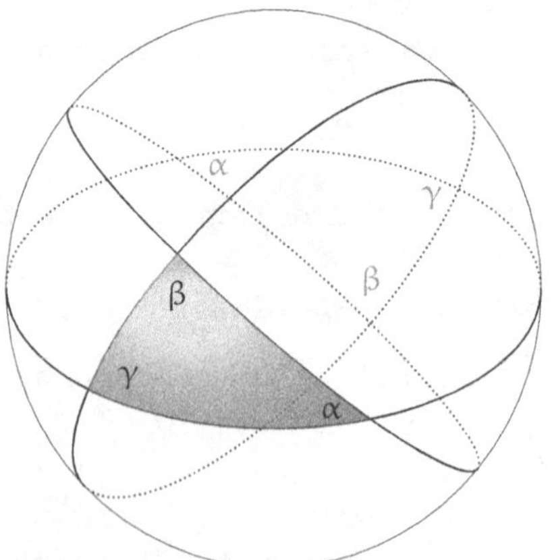

Spherical triangle $\triangle ABC$.

"Sure, but..." I managed.

"You've got two α lunes, two β lunes, and two γ lunes. Six in all, and together they cover the surface of the sphere."

"Okay?" I said, imagining the six lunes coming together to create the sphere.

"The six lunes cover the surface of the sphere, *but*...?" Miruka raised an eyebrow, prompting me to finish her statement.

"But... Oh! There are dupes here! Specifically, there's three lunes overlapping the triangle $\triangle ABC$! So the sphere's surface area will equal the total area of the six lunes, minus the duped parts."

"Right, but don't forget $\triangle ABC$ has a twin on the back side. So there aren't two duplicated areas, there are four."

"Indeed. And that gives us everything we need to set up a formula." I took Miruka's pen back.

$$\underbrace{4\pi R^2}_{\text{sphere's surface area}} = \underbrace{2S_\alpha + 2S_\beta + 2S_\gamma}_{\text{total area of the six lunes}} - \underbrace{4\triangle ABC}_{\text{duped area}}$$

Miruka nodded. "Very good."

I continued with my calculations. The proof I was after was already assembling itself in my mind, I just had to give it form.

$$\begin{aligned}
4\pi R^2 &= 2S_\alpha + 2S_\beta + 2S_\gamma - 4\triangle ABC && \text{from above} \\
4\pi R^2 &= 4\alpha R^2 + 4\beta R^2 + 4\gamma R^2 - 4\triangle ABC && \text{using } S_\alpha = 2\alpha R^2\text{, etc.} \\
\pi R^2 &= \alpha R^2 + \beta R^2 + \gamma R^2 - \triangle ABC && \text{divide both sides by 4} \\
\triangle ABC &= R^2(\alpha + \beta + \gamma - \pi) && \text{rearrange, factor out } R^2
\end{aligned}$$

Answer 8-2 (area of a spherical triangle)

The sphere's surface can be covered by pairs of lunes having angles α, β, γ, with overlap equivalent to four triangles $\triangle ABC$. From this, we obtain

$$4\pi R^2 = 2S_\alpha + 2S_\beta + 2S_\gamma - 4\triangle ABC.$$

From $S_\alpha = 2\alpha R^2, S_\beta = 2\beta R^2, S_\gamma = 2\gamma R^2$,

$$\triangle ABC = R^2(\alpha + \beta + \gamma - \pi).$$

"Done and done," Miruka said.

8.4.4 Gaussian Curvature

"So I didn't need integrals to find the area of a spherical triangle after all," I muttered, slightly miffed. "Now I feel kind of silly, being so excited yesterday when I was thinking about $R \to \infty$ and noticed it was an extension of the sum of interior angles of a triangle equaling 180°."

Miruka gave a slight nod. "If you're going that route, rather than

$$\triangle ABC = R^2 (\alpha + \beta + \gamma - \pi)$$

it's more fun to think about a constant K like

$$K = \frac{\alpha + \beta + \gamma - \pi}{\triangle ABC}.$$

That lets you see the geometry you're using when you make triangles out of geodesics."

- Spherical geometry when $K > 0$
- Euclidean geometry when $K = 0$
- Hyperbolic geometry when $K < 0$

"Whoa," I said. "This one constant K determines everything? It's like... what, a classifier for geometries?"

"You can think of K as representing how far removed from Euclidean geometry you are. Or, you can consider it as describing the 'flatness' of the space you're working in. It's a quantity called Gaussian curvature."

"Gaussian curvature, huh?"

"There are different kinds of curvature. For example, if you let R be the radius of a circle in a plane, you can define the curvature of that circle as $\frac{1}{R}$. The larger the R the smaller the circle's curvature, and vice versa. A big R means the circle is curving gently, while a small R would show a really tight curve, so the reciprocal of R makes for a natural definition. We can also use the curvature of circles to define the curvature at a point on a curve. Roughly put, you can just think of a certain circle touching the curve at that point, and use the curvature $\frac{1}{R}$ of that circle. Or, if it's bending away from the circle, use $-\frac{1}{R}$ instead."

"Interesting. I assume the curvature of a straight line is defined as 0?"

"You assume correctly. To define the curvature of a two-dimensional surface instead, you have to consider how it expands. For example, if you're on a cylinder, in some directions it curves like a circle, but in other directions it's straight like a line, so the amount of curvature depends on which direction you're facing."

"Makes sense."

"Here's how you define the Gaussian curvature at a point P on a curved surface. Consider a plane including P and having a vector tangent to the curved surface at P as a normal vector, and you can find the curvature of the curve formed by the plane and the curved surface. Find the maximum and minimum curvatures when changing the direction of that plane, and their product is the Gaussian curvature at P."

"Okay... Okay, right." I said, tracing through in my head what Miruka had just described.

"Here are some simple examples."

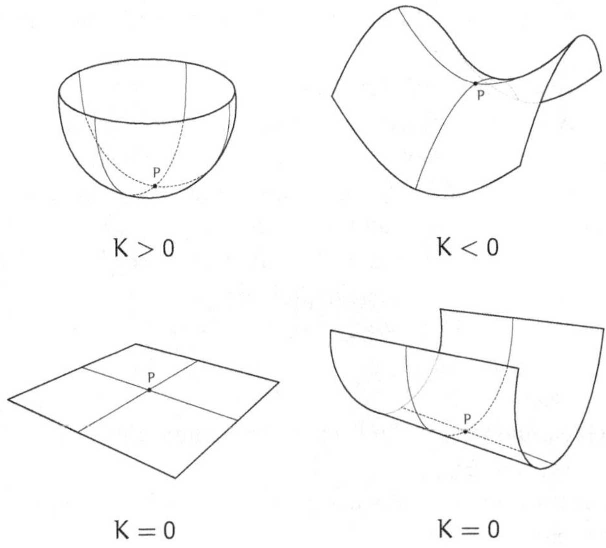

"For each of these curved surfaces, I've drawn the curves with maximum and minimum curvature at P. For a sphere with radius R, the maximum and minimum curvatures will be the same, either $\frac{1}{R}$ or $-\frac{1}{R}$. In either case the signs are the same, so the Gaussian curvature for spherical geometry, in other words $K = \frac{1}{R^2}$, will be positive. For a surface that's curved like a saddle, the maximum and minimum curvatures will have opposite signs, so the Gaussian curvature will be negative, $K = -\frac{1}{R^2}$."

"Got it," I said. "And in the case of a plane the maximum and minimum curvatures are both zero, so $K = 0 \cdot 0 = 0$. So the Gaussian curvature in Euclidean geometry is $K = 0$."

"Exactly. For a cylinder there are two possibilities, a maximum curvature of $\frac{1}{R}$ and a minimum curvature of 0, or a maximum of 0 and a minimum of $-\frac{1}{R}$. In either case one will have a zero value, so their product is 0."

"So a cylinder has a Gaussian curvature of 0? Now that's surprising."

"A surface can curve in different ways, depending on direction. By using the product of the maximum and minimum curvatures, Gaussian curvature considers that change in direction."

"I would have thought you'd take the average of the two."

"Gauss did also talk about mean curvature, which is exactly that."

"Yet another form of curvature!"

"So anyway, for spherical geometry we have $K > 0$, and for Euclidean geometry we have $K = 0$," Miruka continued, her voice suddenly lower. "Since $K = \frac{1}{R^2}$ for a sphere will always be positive this isn't much of a surprise, and if you're thinking about the limit as $R \to \infty$, I guess you could still claim that $K = \frac{1}{R^2}$ for Euclidean geometry too. So what about hyperbolic geometry?"

"Where $K < 0$, right? Well, then $K = \frac{1}{R^2}$ can't be negative, so... Er?"

"If we want to stick to $K = \frac{1}{R^2}$, then $K < 0$ means R is going to be some real-number multiple of the imaginary unit i. So if $K = -1$, for example, $R = \pm i$."

"So now we have a spherical geometry in which radii are imaginary numbers?"

"That's what the math says, right?" Miruka grinned. "If we can think of Euclidean geometry as spherical geometry on an infinitely large sphere, then why not think of hyperbolic geometry as taking place on a sphere with an imaginary radius? This is why it's fun to play with Gaussian curvature."

"Gauss was just... amazing."

"Indeed, but to give credit where credit is due, it was Lambert who, when studying non-Euclidean geometry, made the prophetic discovery that non-Euclidean 'planes' were similar to spheres with imaginary radii."

8.4.5 A Remarkable Theorem

Possibly unable to contain her excitement, Miruka stood up and began walking around me.

"So we talked about the Gaussian curvature of a cylinder," she began. "Specifically, we said that $K = 0$, the same as for a plane. In fact, this has an important meaning. In 1827, when he was studying curved surfaces, Gauss wrote a book called *Disquisitiones generales circa superficies curvas*, or 'General Investigations of Curved Surfaces,' where he first defined what we now call Gaussian curvature as a measure of 'curviness' at some point on a surface. After doing so, he proved a theorem that goes something like this."

> The Gaussian curvature of a surface does not change if one bends the surface without stretching it.

"If you draw a triangle on a piece of paper, the area of that triangle doesn't change when you roll the paper into a cylinder. Same thing if you bend or twist the paper in any other way, so long as you don't stretch or compress it. Its Gaussian curvature doesn't change, either. Invariants like this are extremely important. So long as there's no stretching involved, the Gaussian curvature at a point on a surface is invariant. The Gaussian curvature at a point on a sphere will always be $\frac{1}{R^2}$. The Gaussian curvature at a point on a plane will always be 0. That implies there's no way to turn a sphere into a plane without some stretching."

"So Gaussian curvature serves as an indicator of whether one 'plane' can morph into another?" I said.

"It does, but that's not all. Let's use 'intrinsic' to describe quantities we can obtain by measuring lengths and angles on a curved surface, and 'extrinsic' to describe quantities we can't know without investigating how the curved surface is embedded in a space. The Gaussian curvature K is defined using extrinsic quantities, but Gauss himself provided calculations showing that K can be represented using intrinsic quantities. In fact, Gaussian curvature was the first intrinsic quantity to be discovered.

"If you imagine a plane in three-dimensional space and that plane rolled up into a cylinder, they look like different surfaces. But actually, we can consider them as only differing according to how they are embedded in the three-dimensional space. No matter how you twist or bend a plane in a three-dimensional space, the Gaussian curvature at each point will remain unchanged. We can think of Gaussian curvature as a quantity describing how much a curved surface bends, regardless of the space in which it is embedded."

"I'm . . . not sure I understand what that means," I admitted.

"Think of it like this. A two-dimensional being can calculate Gaussian curvatures just by measuring lengths and angles—intrinsic properties of the plane it lives in. It doesn't have to consider the 'outside' of its space at all. That's quite remarkable, considering that Gaussian curvature is defined using *extrinsic* properties. In fact, Gauss called this his *Theorema Egregium*, his 'Remarkable Theorem.'"

8.4.6 Homogeneity and Isotropy

"The intrinsic properties of Gaussian curvature had a huge impact on geometry," Miruka continued. "The mathematician Riemann generalized the Gaussian curvature of surfaces to consider curvature of n-dimensional spaces. This generalization can take various forms. For spherical, Euclidean, and hyperbolic geometry, Gaussian curvature has a constant value K. This condition that K is constant is called homogeneity. It means the Gaussian curvature does not depend on location within the space.

"We can generalize by getting rid of this presumption of homogeneity. In other words, the Gaussian curvature K depends on the point p in the space. In that case, Gaussian curvature becomes some function $K(p)$, so the formula for the area of the $\triangle ABC$ you were after isn't a product, it's an integral. Along with an extension by Bonnet, this is called the Gauss–Bonnet theorem."

$$\alpha + \beta + \gamma - \pi = K\Delta(\triangle ABC) \quad \text{Gaussian curvature a constant } K$$

$$\alpha + \beta + \gamma - \pi = \iint_{\triangle ABC} K(p)\,dS \quad \text{Gaussian curvature a function } K(p)$$

"If this function $K(p)$ gives the Gaussian curvature, then so long as you know where on the surface you are, you can figure out the Gaussian curvature there. There's room for generalization here, too, given the assumption that the curvature doesn't change depending on which way you're facing. That property is called isotropy.

"In higher dimensions, we can remove the isotropy condition and think of even more generalized curvature. You end up treating the curvature at p not as a real number like Gaussian curvature, but as something called a 'curvature tensor' instead. Unfortunately, I don't know enough about those to comfortably talk about them. I need to study more."

Miruka glanced at me, blushing. Just then, the bell rang.

8.4.7 In Return

"I need to get home," Miruka said, collecting the papers covering the table.

"Yeah, me too," I said. We were the last two students in the lounge.

"So how's your vocabulary list coming?"

"My vocabulary list?"

"Spiteful, smug, ruminative, aloof, detached... Any additions?"

"Yeah, I've been wanting to apologize for that," I said, looking down in embarrassment. "Seriously, I'm sorry. I have no excuse for taking out my frustrations on you."

"Did you get partial credit?"

"I did, for forgetting the variable substitution. I'm studying to shore up some weak areas. My literature score was... pretty bad. I need to doing something about that before my next practice test."

"Hmph."

"Anyway, I said some awful things. I hope you'll forgive me."

"It wasn't so awful," Miruka said. "It was an interesting point of view. Here's your reward. Hold out your hand."

Confused as I was, I held out a hand as instructed. Miruka pulled a small bag out of her backpack, and placed it on my palm.

"What's this? Chocolate?" *Forgiveness?*

"Smell it."

She took a step toward me and opened the bag, releasing a scent of cacao. She move closer still, bringing her face closer to mine. I looked into her eyes, and she looked into mine.

"Do you have plans for Christmas eve?" she suddenly asked.

"Just studying. Holidays don't mean much for me just now."

"There's going to be a seminar. Like the one last year, when the theme was Fermat's last theorem."

"I don't think I'll have time. That's right after my next exam, and I'll—"

"It's just a few hours."

"Maybe if I get an A ranking on that test . . . "

"Then you'd better get that A." Miruka snapped her fingers. "You need to study more."

> This measure of how a curved surface is curving, discovered by Gauss in 1827 and now called Gaussian curvature, was the first demonstration of an intrinsic property of curved surfaces. . . . It strongly suggested that we can talk about the curvature of our universe without there being something "outside" of it.
>
> Toshikazu Sunada
> *The Geometry of Curved Surfaces*

Chapter 9

Inspiration and Perspiration

> When we arrived at Coutances, we got into a brake to go for a drive, and, just as I put my foot on the step, the idea came to me, though nothing in my former thoughts seemed to have prepared me for it, that the transformations I had used to define Fuchsian functions were identical with those of non-Euclideian geometry.
>
> Henri Poincaré
> *Science and Method*

9.1 Trigonometric Training

9.1.1 Inspiration Versus Perspiration

Many people consider mathematics to be a field dominated by inspiration. Indeed, books describing the history of math are filled with moments of inspiration. These sudden flashes of insight make for some thrilling drama, and sometimes even change the world. For some, this is what mathematics is all about.

Sure, I suppose we wouldn't have today's mathematics without inspiration. Even so, I also suspect that equally important is the unimaginable amount of toil required for that inspiration to manifest in the first place. Then, having achieved some burst of revelation, mathematicians would be faced with still more mountains to climb: verification, formulation, extension, generalization...

Things like that would be flitting about the edges of my consciousness when I found myself deep in some involved calculation. Moving terms, expanding terms, differentiating, integrating, substituting... I couldn't arrive at a solution without going through long sequences of these boring, everyday operations. Not that I could compare my college entrance exam studies with the inspiration of geniuses, but still.

Occasionally I would come across problems that could be solved with just a quick flash of insight. Geometry proofs that could be completed by drawing a line at exactly the right place, or algebra problems that practically solved themselves once I'd grasped their overall form. But those were exceptions. In most cases, even when it was clear how to solve a problem, I couldn't reach its solution without a lot of careful work. It felt good to quickly solve a problem through an "ah-ha!" moment, but I couldn't always count on that; persistence is at least as important. Inspiration and perspiration are both needed.

"You need to study more," Miruka had said. As if I needed that advice. Of course I would study more; that was my job. Just like an athlete has to always keep training, I had to keep on practicing. I at least needed to polish my skills until I could perform the basics with certainty. One of my core exercises was something I called "trigonometric training." For example...

9.1.2 The Unit Circle

Trigonometric functions are circle functions.

We can define $\cos\theta$ and $\sin\theta$ using the unit circle, a circle with radius 1 centered at the origin. Letting (x, y) be a point on the unit circle, and letting θ be the angle between the positive side of the x-axis and a line connecting (x, y) and the origin, $\cos\theta$ and $\sin\theta$ are defined like this.

$$\begin{cases} \cos\theta = x \\ \sin\theta = y \end{cases}$$

In other words, $\cos\theta$ is the x-coordinate of the point, and $\sin\theta$ is the y-coordinate. The equation describing the unit circle is $x^2 + y^2 = 1^2$, which we can now rewrite as

$$\cos^2\theta + \sin^2\theta = 1^2.$$

These are the absolute basics.

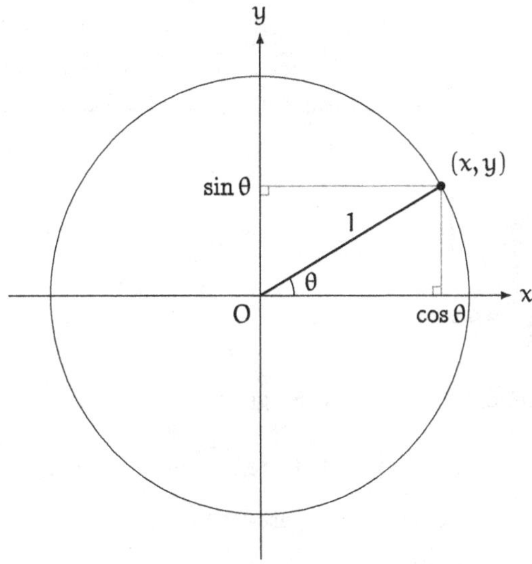

The unit circle.

The location of a point won't change when we increase θ by 2π, so

$$\begin{cases} \cos\theta = \cos(\theta + 2\pi) = \cos(\theta - 2\pi) \\ \sin\theta = \sin(\theta + 2\pi) = \sin(\theta - 2\pi) \end{cases}.$$

Letting n be an integer, we can generalize this as

$$\begin{cases} \cos\theta = \cos(\theta + 2n\pi) \\ \sin\theta = \sin(\theta + 2n\pi) \end{cases}.$$

Here's what happens to θ in the cases where the x- and y-coordinates are 0.

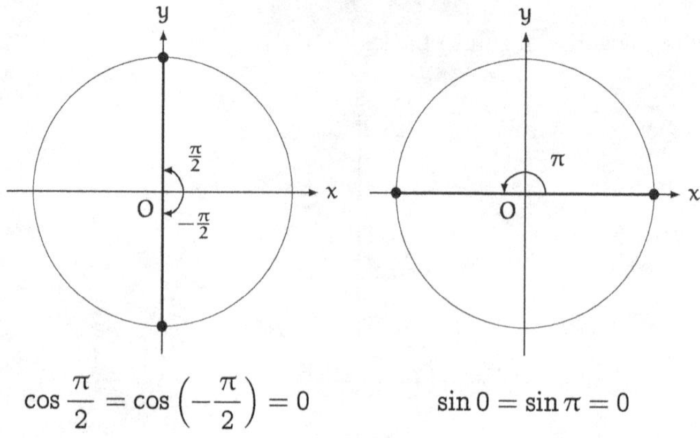

$$\cos\frac{\pi}{2} = \cos\left(-\frac{\pi}{2}\right) = 0 \qquad\qquad \sin 0 = \sin\pi = 0$$

Again, generalizing for an integer n...

$$\begin{cases} \cos\left(n\pi + \dfrac{\pi}{2}\right) = 0 & \text{(integer multiple of } \pi) + \dfrac{\pi}{2} \\ \sin n\pi = 0 & \text{integer multiple of } \pi \end{cases}$$

If you think about the unit circle,

$$\sin\,(\text{integer multiple of } \pi) = 0$$

makes perfect sense, because when θ is an integer multiple of π, the point will always be on the x-axis. The sine value is the y-coordinate, so of course $\sin\theta$ will be 0.

Considering the range of values the point's x- and y-coordinates can take, we can see that $\cos\theta$ and $\sin\theta$ will each be at least -1 and at most 1.

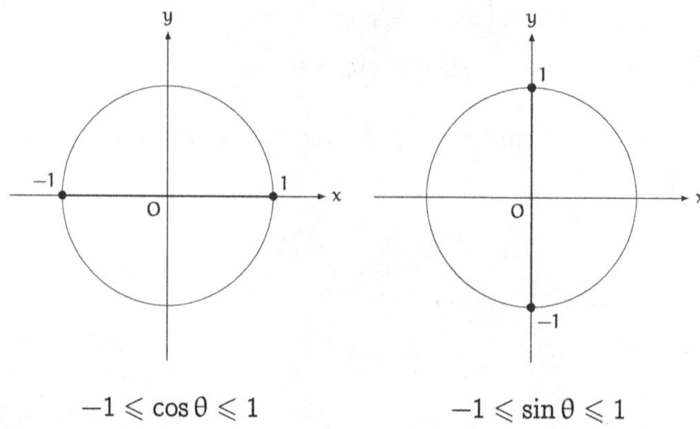

$$-1 \leqslant \cos\theta \leqslant 1 \qquad\qquad -1 \leqslant \sin\theta \leqslant 1$$

So let's think about the conditions required for these inequalities to hold true. From the θ values that make the x-coordinate 1 or -1, we get

$$\begin{cases} \cos 0 = 1 \\ \cos \pi = -1 \end{cases},$$

or, more generally,

$$\begin{cases} \cos(2n\pi + 0) = 1 & \text{an even multiple of } \pi \\ \cos(2n\pi + \pi) = -1 & \text{an odd multiple of } \pi \end{cases}.$$

In other words,

$$\begin{cases} \cos(\text{an even multiple of } \pi) = 1 \\ \cos(\text{an odd multiple of } \pi) = -1 \end{cases}.$$

Together, we get

$$\cos(\text{the } n\text{-th multiple of } \pi) = (-1)^n.$$

Conversely, from the θ values that make the y-coordinate 1 or -1, we get

$$\begin{cases} \sin \dfrac{\pi}{2} = 1 \\ \sin\left(-\dfrac{\pi}{2}\right) = -1 \end{cases},$$

which we can generalize as

$$\begin{cases} \sin\left(2n\pi + \dfrac{\pi}{2}\right) = 1 \\ \sin\left(2n\pi - \dfrac{\pi}{2}\right) = -1 \end{cases}.$$

9.1.3 The Sine Curve

Graphs of $x = \cos\theta$ and $y = \sin\theta$ look like this.

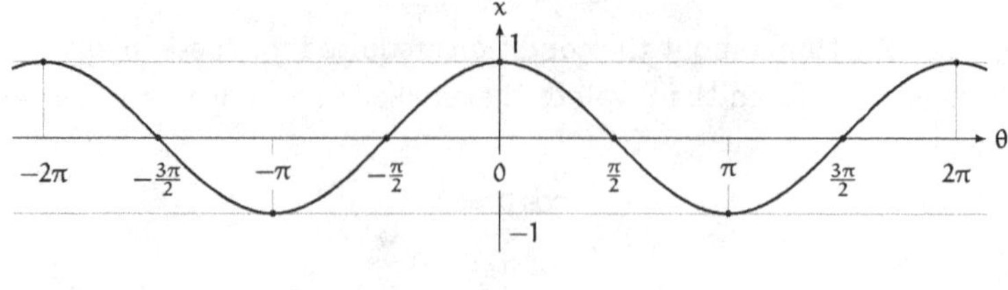

$x = \cos\theta$

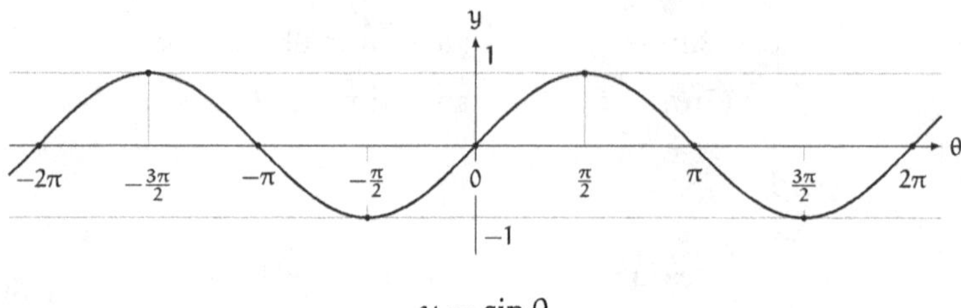

$y = \sin\theta$

To create the graph of $\sin\theta$ from the graph of $\cos\theta$, you just move it $\frac{\pi}{2}$ horizontally. It's easy to get mixed up about whether to add or subtract $\frac{\pi}{2}$, so you gotta be careful.

$$\begin{cases} \cos\left(\theta - \dfrac{\pi}{2}\right) = \sin\theta \\ \sin\left(\theta + \dfrac{\pi}{2}\right) = \cos\theta \end{cases}$$

From symmetry, we can see that this holds, too.

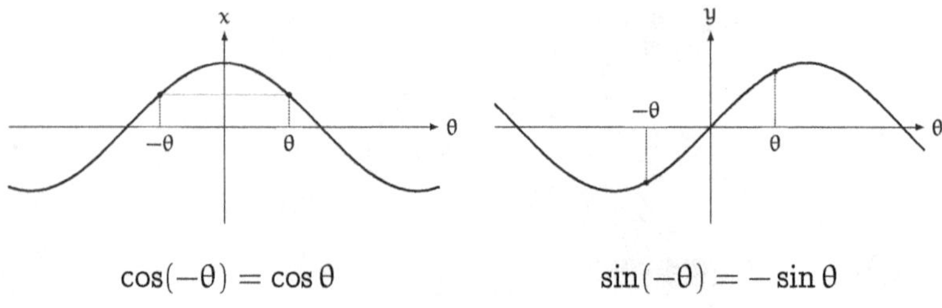

$\cos(-\theta) = \cos\theta$ $\qquad$ $\sin(-\theta) = -\sin\theta$

So $\cos\theta$ is an even function and $\sin\theta$ is odd.

9.1.4 *Rotation Matrix to Angle Sum Identities*

The point (u, v) we get by rotating (x, y) by θ degrees counterclockwise around the origin is

$$\begin{cases} u = x\cos\theta - y\sin\theta \\ v = x\sin\theta + y\cos\theta \end{cases}.$$

We can also write this as the rotation matrix

$$\begin{pmatrix} u \\ v \end{pmatrix} = \begin{pmatrix} \cos\theta & -\sin\theta \\ \sin\theta & \cos\theta \end{pmatrix} \begin{pmatrix} x \\ y \end{pmatrix}.$$

We can think of a rotation by $(\alpha+\beta)$ as a rotation by α followed by a rotation by β. This is a product of rotation matrices.

$$\begin{aligned}
&\begin{pmatrix} \cos(\alpha+\beta) & -\sin(\alpha+\beta) \\ \sin(\alpha+\beta) & \cos(\alpha+\beta) \end{pmatrix} \\
&= \begin{pmatrix} \cos\beta & -\sin\beta \\ \sin\beta & \cos\beta \end{pmatrix} \begin{pmatrix} \cos\alpha & -\sin\alpha \\ \sin\alpha & \cos\alpha \end{pmatrix} \\
&= \begin{pmatrix} \cos\beta\cos\alpha - \sin\beta\sin\alpha & -\cos\beta\sin\alpha - \sin\beta\cos\alpha \\ \sin\beta\cos\alpha + \cos\beta\sin\alpha & -\sin\beta\sin\alpha + \cos\beta\cos\alpha \end{pmatrix} \\
&= \begin{pmatrix} \cos\alpha\cos\beta - \sin\alpha\sin\beta & -(\sin\alpha\cos\beta + \cos\alpha\sin\beta) \\ \sin\alpha\cos\beta + \cos\alpha\sin\beta & \cos\alpha\cos\beta - \sin\alpha\sin\beta \end{pmatrix}
\end{aligned}$$

We obtain the angle sum identities by comparing elements here.

$$\begin{cases} \cos(\alpha+\beta) = \cos\alpha\,\cos\beta - \sin\alpha\,\sin\beta \\ \sin(\alpha+\beta) = \sin\alpha\,\cos\beta + \cos\alpha\,\sin\beta \end{cases}$$

9.1.5 *Angle Sum Identities to Product-to-Sum Identities*

There are also angle difference identities, but the angle sum identities

$$\begin{cases} \cos(\alpha+\beta) = \cos\alpha\,\cos\beta - \sin\alpha\,\sin\beta \\ \sin(\alpha+\beta) = \sin\alpha\,\cos\beta + \cos\alpha\,\sin\beta \end{cases}$$

are all you need. This is because when you have a difference $\alpha - \beta$, you can just think of it as $\alpha + (-\beta)$ instead. Using $\cos(-\beta) = \cos\beta$ and $\sin(-\beta) = -\sin\beta$, we get

$$\begin{cases} \cos(\alpha - \beta) = \cos\alpha\,\cos(-\beta) - \sin\alpha\,\sin(-\beta) \\ \qquad\qquad\;\; = \cos\alpha\,\cos\beta + \sin\alpha\,\sin\beta \\ \sin(\alpha - \beta) = \sin\alpha\,\cos(-\beta) + \cos\alpha\,\sin(-\beta) \\ \qquad\qquad\;\; = \sin\alpha\,\cos\beta - \cos\alpha\,\sin\beta \end{cases}.$$

Considering the angle sum identities in the case where $\theta = \alpha = \beta$, we can derive the double angle formulas,

$$\begin{cases} \cos 2\theta = \cos^2\theta - \sin^2\theta \\ \sin 2\theta = 2\sin\theta\cos\theta \end{cases}.$$

Using $\cos^2\theta + \sin^2\theta = 1$, we can write $\cos 2\theta$ as

$$\begin{cases} \cos 2\theta = 1 - 2\sin^2\theta & \text{using } \cos^2\theta = 1 - \sin^2\theta \\ \cos 2\theta = 2\cos^2\theta - 1 & \text{using } \sin^2\theta = 1 - \cos^2\theta \end{cases}.$$

Solving for $\sin^2\theta$ and $\cos^2\theta$, we can convert squares to double angles.

$$\begin{cases} \sin^2\theta = \dfrac{1}{2}(1 - \cos 2\theta) \\ \cos^2\theta = \dfrac{1}{2}(1 + \cos 2\theta) \end{cases}$$

These are called the half-angle identities when written like this.

$$\begin{cases} \sin^2\dfrac{\theta}{2} = \dfrac{1}{2}(1 - \cos\theta) \\ \cos^2\dfrac{\theta}{2} = \dfrac{1}{2}(1 + \cos\theta) \end{cases}$$

The angle sum identities we derived earlier looked like this.

$$\begin{cases} \cos(\alpha + \beta) = \cos\alpha\,\cos\beta - \sin\alpha\,\sin\beta & \cdots ① \\ \sin(\alpha + \beta) = \sin\alpha\,\cos\beta + \cos\alpha\,\sin\beta & \cdots ② \\ \cos(\alpha - \beta) = \cos\alpha\,\cos\beta + \sin\alpha\,\sin\beta & \cdots ③ \\ \sin(\alpha - \beta) = \sin\alpha\,\cos\beta - \cos\alpha\,\sin\beta & \cdots ④ \end{cases}$$

From this we can derive the product-to-sum identities.

$$\begin{cases} \cos\alpha\cos\beta = \frac{1}{2}\Big(\cos(\alpha+\beta)+\cos(\alpha-\beta)\Big) & \text{from } \frac{1}{2}(①+③) \\ \sin\alpha\sin\beta = -\frac{1}{2}\Big(\cos(\alpha+\beta)-\cos(\alpha-\beta)\Big) & \text{from } -\frac{1}{2}(①-③) \\ \sin\alpha\cos\beta = \frac{1}{2}\Big(\sin(\alpha+\beta)+\sin(\alpha-\beta)\Big) & \text{from } \frac{1}{2}(②+④) \\ \cos\alpha\sin\beta = \frac{1}{2}\Big(\sin(\alpha+\beta)-\sin(\alpha-\beta)\Big) & \text{from } \frac{1}{2}(②-④) \end{cases}$$

Memorizing formulas has its place, but if you don't know how to use them, what's the point? Deriving all these identities in my "trigonometric training" regimen allowed me to practice how they're actually used. Also, knowing I could derive these identities on my own meant I didn't need to worry about forgetting them.

9.1.6 Mom

"Are you even listening to me?" my mother said.

I had been doing some trigonometric training in my head while washing dishes after dinner, but her words brought me back to reality.

"Yeah, sure," I said. "Glad to hear you're feeling better."

"That's *not* what I was talking about!" She poked me in the side with her elbow while drying off a dish. "But yes, I am feeling better, thank you."

"Well that's good. I guess Dad will be late again tonight?"

"You know how busy he gets at year-end. And how's your work coming along?"

"Okay, I guess. You sound like Miruka, by the way."

"Do I now?" This apparently made her quite happy. "You should be thankful you have such good friends."

"What brought that on?"

"Your father, I guess."

"Dad? How so?"

"From when he quit his job."

"Wait, what? He quit his job? Seriously?"

"Oh, this was long ago, right after we got married. He was working at a company where he didn't really have any friends. He was so tired when he came home every day. Then, one day he just up and quit. He wasn't working at all for a time. I never told you about this?"

"No, never." Honestly, I'd never even considered what my parents were like when they were newlyweds.

"So instead of going to work every day, he went fishing."

"Fishing? *Dad* went fishing?"

"Him and me both, every day, from morning to night. I'd fill a canteen with tea and pack some rice balls, and just sit there while he fished. I remember it was still cold enough to need a coat when he started, and that lasted until spring, around when the cherry blossoms had all fallen."

"Wow."

"He stopped going just as suddenly as he'd started. I tried to bait his hook for him, and cut my finger pretty bad. See this scar here? There was blood everywhere, all over my clothes... He started looking for work the very next day."

"Huh. I had no idea..."

She chuckled. "There's still a lot you don't know."

The only image I had of my parents was exactly that—them being my parents—but that was just an illusion. They'd of course each had their own lives before I was around, lives I knew nothing about. Still, it was difficult to imagine my jobless father going fishing, my mother eating rice balls under a nearby cherry tree.

"I guess it's important to have someone near to you," I muttered, unintentionally vocalizing my thoughts.

"Well of course it is. But 'near' doesn't have to refer to distance."

"What do you mean?"

"Never mind that. I'm good here, you go take a bath. You have a practice exam tomorrow, right?"

9.2 The Practice Exam

9.2.1 Keeping Cool

Indeed, I did have another practice exam the next day.

I showed up early enough to use the toilet and do some light stretching exercises. After entering the room and sitting down, I placed my pencils and such on the desk along with a small clock, ensuring its alarm was turned off. With everything in place, once the test started I made sure to completely focus my attention on the problems before me.

This was my last practice test. The last time I would come to this testing center, the last time I'd be here with these other high-school students, the last time I would have to suffer this "noisy silence." The last time until the real thing, at least.

I was living on test time. I'd gone to bed the night before and woken up this morning right on schedule. But even as prepared as I felt, it took a lot of composure to keep from losing my cool. At the same time, it took a certain amount of passion to remain engaged with the test problems. A sort of heated composure, if you will.

I was absolutely determined to get an "A" ranking on that test. It was my one job. I had put a lot of effort into making that a possibility, and I was sure my hard work would pay off.

Please, please pay off...

The bell rang, and everyone simultaneously opened their test booklets. Time to get to work.

9.2.2 Avoiding Pitfalls

Problem 9-1 (integrals of trigonometric functions)

Find the integral

$$\int_{-\pi}^{\pi} \sin mx \sin nx \, dx$$

for positive integers m and n.

The form of this problem showed me exactly where I needed to go. I had a product of two functions in x, $\sin mx$ and $\sin nx$. Products make integrals tricky to deal with; the easier approach would be to convert these to sums. Time to put my trigonometric training to work.

$$\sin\alpha \sin\beta = -\frac{1}{2}\Big(\cos(\alpha+\beta) - \cos(\alpha-\beta)\Big) \quad \text{product-to-sum formula}$$

Letting $\alpha = mx$ and $\beta = nx$, I could use this to rewrite the problem using sums.

$$\begin{aligned}\sin mx \sin nx &\\ &= -\frac{1}{2}\Big(\cos(mx+nx) - \cos(mx-nx)\Big)\\ &= -\frac{1}{2}\Big(\cos(m+n)x - \cos(m-n)x\Big) \qquad \text{factor out an } x\end{aligned}$$

Now that I had a sum, I could proceed with the integral.

$$\begin{aligned}\int_{-\pi}^{\pi} \sin mx \sin nx \, dx &= -\frac{1}{2}\int_{-\pi}^{\pi}\Big(\cos(m+n)x - \cos(m-n)x\Big)\,dx\\ &= \underbrace{-\frac{1}{2}\int_{-\pi}^{\pi}\cos(m+n)x\,dx}_{①} + \underbrace{\frac{1}{2}\int_{-\pi}^{\pi}\cos(m-n)x\,dx}_{②}\end{aligned}$$

Now I needed to find two integrals, ① and ②. The first was easy enough, because when $x = \pm\pi$, the $\sin(m+n)x$ that appears in the integral equals 0.

$$\begin{aligned}\int_{-\pi}^{\pi}\cos(m+n)x\,dx &= \frac{1}{m+n}\Big[\sin(m+n)x\Big]_{-\pi}^{\pi} \qquad \text{take the integral}\\ &= 0 \qquad \text{because } \sin(\text{int. mult. of } \pi) = 0\end{aligned}$$

The pitfall here was jumping to the conclusion that I could do the same thing for ②, which had $m - n$. I had to consider two possibilities here, the cases where $m - n \neq 0$ and where $m - n = 0$. No way I was going to lose points by trying to divide by zero! *Keep it cool, keep it cool...*

② would be the same as ① when $m - n \neq 0$.

$$\begin{aligned}\int_{-\pi}^{\pi}\cos(m-n)x\,dx &= \frac{1}{m-n}\Big[\sin(m-n)x\Big]_{-\pi}^{\pi} \qquad \text{take the integral}\\ &= 0 \qquad \text{because } \sin(\text{int. mult. of } \pi) = 0\end{aligned}$$

But in ②, a $\cos 0x$ appears when $m - n = 0$, and this of course equals 1, since $\cos 0 = 1$.

$$\begin{aligned}
\int_{-\pi}^{\pi} \cos(m-n)x\,dx &= \int_{-\pi}^{\pi} \cos 0x\,dx && \text{because } m-n=0 \\
&= \int_{-\pi}^{\pi} 1\,dx && \text{because } \cos 0x = \cos 0 = 1 \\
&= \Big[x\Big]_{-\pi}^{\pi} && \text{take the integral} \\
&= \pi - (-\pi) \\
&= 2\pi
\end{aligned}$$

All that's left is some careful addition.

$$\int_{-\pi}^{\pi} \sin mx \sin nx\,dx = -\frac{1}{2}\underbrace{\int_{-\pi}^{\pi} \cos(m+n)x\,dx}_{①} + \frac{1}{2}\underbrace{\int_{-\pi}^{\pi} \cos(m-n)x\,dx}_{②}$$

① would equal 0, regardless of the values of m and n. ② would equal 0 when $m - n \neq 0$, or 2π when $m - n = 0$. So now, being careful of the $\frac{1}{2}$ coefficient...

$$\int_{-\pi}^{\pi} \sin mx \sin nx\,dx = \begin{cases} 0 & \text{when } m-n \neq 0 \\ \pi & \text{when } m-n = 0 \end{cases}$$

As Miruka would say, done and done.

Answer 9-1 (integrals of trigonometric functions)

For integers m and n,

$$\int_{-\pi}^{\pi} \sin mx \sin nx\,dx = \begin{cases} 0 & \text{when } m-n \neq 0 \\ \pi & \text{when } m-n = 0 \end{cases}.$$

Doing good so far. Keeping my cool. Even better, a heated composure.

On to the next problem...

9.2.3 Inspiration or Perspiration?

Problem 9-2 (definite integral with parameters)

Consider a definite integral with real-number parameters a and b,

$$I(a,b) = \int_{-\pi}^{\pi} \left(a + b\cos x - x^2\right)^2 dx.$$

Find the minimum value for $I(a,b)$ and the corresponding values for a and b.

I figured the procedure I should follow would be something like this.

Step 1: Expand $(a + b\cos x - x^2)^2$ to create a sum.

Step 2: Find the definite integral $I(a,b)$.

Step 3: Find the minimum value of $I(a,b)$.

Step 1 would just be some simple algebra, and the results of Step 2 should be a quadratic involving a and b. Then I could just complete the square in Step 3. Nothing difficult here, and no need for inspiration. Of course, that expansion would likely produce a lot of terms, so there might be some perspiration involved.

I paused, wondering if I should stick with this, or if it would be better to move on to another problem first. After a few seconds, however, I decided to press on. If I could maintain my heated composure and proceed carefully, I should be able to tackle this.

Onward, then!

Step 1, then. I wanted to expand $(a + b\cos x - x^2)^2$ into a sum.

$$\begin{aligned} I(a,b) &= \int_{-\pi}^{\pi} \left(a + b\cos x - x^2\right)^2 dx \\ &= \int_{-\pi}^{\pi} \Big(\underbrace{a^2}_{①} + \underbrace{b^2\cos^2 x}_{②} + \underbrace{x^4}_{③} + \underbrace{2ab\cos x}_{④} - \underbrace{2bx^2\cos x}_{⑤} - \underbrace{2ax^2}_{⑥}\Big)\, dx \end{aligned}$$

Okay, on to Step 2, calculating the definite integral $I(a,b)$.

Now I needed to find the individual integrals for each of the terms ① through ⑥. The first was a constant a^2, so that wasn't hard.

$$\begin{aligned}\int_{-\pi}^{\pi} a^2\,dx &= a^2\Big[\,x\,\Big]_{-\pi}^{\pi} && \text{take the integral}\\ &= a^2\,(\pi-(-\pi))\\ &= 2\pi a^2 \qquad \cdots ①'\end{aligned}$$

For ②, I wanted to get rid of those squares. Not a problem, since by using $\cos^2 x = \frac{1}{2}\,(1+\cos 2x)$, I could rewrite the squares as double angles.

$$\begin{aligned}\int_{-\pi}^{\pi} b^2\cos^2 x\,dx &= b^2\int_{-\pi}^{\pi}\cos^2 x\,dx\\ &= b^2\int_{-\pi}^{\pi}\frac{1}{2}\,(1+\cos 2x)\,dx && \text{squares to double angles}\\ &= \frac{b^2}{2}\int_{-\pi}^{\pi}(1+\cos 2x)\,dx\\ &= \frac{b^2}{2}\Big[\,x+\frac{1}{2}\sin 2x\,\Big]_{-\pi}^{\pi} && \text{take the integral}\\ &= \frac{b^2}{2}\,(\pi-(-\pi)) && \text{because } \sin(\text{integer multiple of }\pi) = 0\\ &= \pi b^2 \qquad \cdots ②'\end{aligned}$$

③ was the integral of x^4, a straightforward calculation.

$$\begin{aligned}\int_{-\pi}^{\pi} x^4\,dx &= \frac{1}{5}\Big[\,x^5\,\Big]_{-\pi}^{\pi} && \text{take the integral}\\ &= \frac{1}{5}\left(\pi^5-(-\pi)^5\right)\\ &= \frac{2\pi^5}{5} \qquad \cdots ③'\end{aligned}$$

④ was the integral of $\cos x$. Again, simple.

$$\begin{aligned}\int_{-\pi}^{\pi} 2ab\cos x\,dx &= 2ab\Big[\,\sin x\,\Big]_{-\pi}^{\pi} && \text{take the integral}\\ &= 2ab\,(0-0) && \text{because } \sin(\text{integer multiple of }\pi) = 0\\ &= 0 \qquad \cdots ④'\end{aligned}$$

⑤ is where I first had to pause and think. *Let's see, the integral of* $x^2 \cos x$ *would be... Ah, right. Integration by parts.*

I could set aside the 2b coefficient for now, and first clean up the $x^2 \cos x$.

$$\int_{-\pi}^{\pi} x^2 \cos x \, dx = \int_{-\pi}^{\pi} x^2 (\sin x)' \, dx \qquad \text{because } \cos x = (\sin x)'$$

$$= \left[x^2 \sin x \right]_{-\pi}^{\pi} - \int_{-\pi}^{\pi} (x^2)' \sin x \, dx \qquad \text{integration by parts}$$

$$= (0 - 0) - \int_{-\pi}^{\pi} (x^2)' \sin x \, dx \qquad \text{because } \sin(\text{integer multiple of } \pi) = 0$$

$$= -\int_{-\pi}^{\pi} 2x \sin x \, dx \qquad \text{because } (x^2)' = 2x$$

$$= -2 \int_{-\pi}^{\pi} x \sin x \, dx \qquad \int_{-\pi}^{\pi} x \sin x \, dx \text{ remains} \ldots$$

Once more, to take care of that $\int_{-\pi}^{\pi} x \sin x \, dx$.

$$\int_{-\pi}^{\pi} x \sin x \, dx$$

$$= \int_{-\pi}^{\pi} x(-\cos x)' \, dx \qquad \text{because } \sin x = (-\cos x)'$$

$$= \left[x(-\cos x) \right]_{-\pi}^{\pi} - \int_{-\pi}^{\pi} (x)'(-\cos x) \, dx \qquad \text{integration by parts}$$

$$= 2\pi + \int_{-\pi}^{\pi} \cos x \, dx \qquad \text{because } -\cos \pi = -\cos(-\pi) = 1$$

$$= 2\pi + \left[\sin x \right]_{-\pi}^{\pi} \qquad \text{take integral}$$

$$= 2\pi \qquad \text{because } \sin(\text{integer multiple of } \pi) = 0$$

And that takes care of the $x^2 \cos x$.

$$\int_{-\pi}^{\pi} x^2 \cos x \, dx = -2 \int_{-\pi}^{\pi} x \sin x \, dx$$

$$= -2 \cdot 2\pi$$

$$= -4\pi$$

Oops, can't forget about that 2b *I set aside earlier!*

$$2b\int_{-\pi}^{\pi} x^2\cos x\,dx = 2b\cdot(-4\pi)$$
$$= \boxed{-8\pi b} \qquad \cdots ⑤'$$

⑥ was x^2, another easy one.

$$2a\int_{-\pi}^{\pi} x^2\,dx = \frac{2a}{3}\Big[x^3\Big]_{-\pi}^{\pi}$$
$$= \frac{2a}{3}\left(\pi^3 - (-\pi)^3\right)$$
$$= \boxed{\frac{4\pi^3}{3}a} \qquad \cdots ⑥'$$

Okay, time to add it all up.

$$I(a,b) = \int_{-\pi}^{\pi}(\underbrace{a^2}_{①} + \underbrace{b^2\cos^2 x}_{②} + \underbrace{x^4}_{③} + \underbrace{2ab\cos x}_{④} - \underbrace{2bx^2\cos x}_{⑤} - \underbrace{2ax^2}_{⑥})\,dx$$
$$= \underbrace{2\pi a^2}_{①'} + \underbrace{\pi b^2}_{②'} + \underbrace{\frac{2\pi^5}{5}}_{③'} + \underbrace{0}_{④'} - \underbrace{(-8\pi b)}_{⑤'} - \underbrace{\frac{4\pi^3}{3}a}_{⑥'}$$
$$= 2\pi a^2 - \frac{4\pi^3}{3}a + \pi b^2 + 8\pi b + \frac{2\pi^5}{5}$$

Now clean up for a and b.

$$= 2\pi\underbrace{\left(a^2 - \frac{2\pi^2}{3}a\right)}_{Ⓐ} + \pi\underbrace{\left(b^2 + 8b\right)}_{Ⓑ} + \frac{2\pi^5}{5}$$

Okay, ready for Step 3, finding the minimum value of $I(a,b)$. To do that, I needed to complete the square for both a and b.

$$\begin{aligned}
\textcircled{A} &= a^2 - \frac{2\pi^2}{3}a \\
&= \left(a - \frac{\pi^2}{3}\right)^2 - \left(\frac{\pi^2}{3}\right)^2 && \text{complete the square} \\
&= \left(a - \frac{\pi^2}{3}\right)^2 - \frac{\pi^4}{9}
\end{aligned}$$

$$\begin{aligned}
\textcircled{B} &= b^2 + 8b \\
&= (b+4)^2 - 4^2 && \text{complete the square} \\
&= (b+4)^2 - 16
\end{aligned}$$

From that, I could represent $I(a, b)$ like this.

$$\begin{aligned}
I(a,b) &= 2\pi \times \textcircled{A} + \pi \times \textcircled{B} + \frac{2\pi^5}{5} \\
&= 2\pi\left\{\left(a - \frac{\pi^2}{3}\right)^2 - \frac{\pi^4}{9}\right\} + \pi\left\{(b+4)^2 - 16\right\} + \frac{2\pi^5}{5} \\
&= 2\pi\left(a - \frac{\pi^2}{3}\right)^2 - \frac{2\pi^5}{9} + \pi(b+4)^2 - 16\pi + \frac{2\pi^5}{5} \\
&= 2\pi\left(a - \frac{\pi^2}{3}\right)^2 + \pi(b+4)^2 - \frac{2\pi^5}{9} + \frac{2\pi^5}{5} - 16\pi \\
&= \underline{2\pi\left(a - \frac{\pi^2}{3}\right)^2} + \underline{\pi(b+4)^2} + \frac{8\pi^5}{45} - 16\pi
\end{aligned}$$

The underlined parts would each be at least 0, so $I(a, b)$ would have a minimum value when both equal zero. Thus, it would take a minimum value when $a = \frac{\pi^2}{3}$ and $b = -4$, and that value would be

$$\frac{8\pi^5}{45} - 16\pi.$$

And there's my answer!

Answer 9-2 (definite integral with parameters)

The definite integral $I(a, b)$ takes a minimum value of $\frac{8\pi^5}{45} - 16\pi$ when $a = \frac{\pi^2}{3}$ and $b = -4$.

Not too much perspiration after all...
Okay, on to the next one!

9.3 Seeing Form

9.3.1 Reading Probability Density Functions

The next day, I felt good about how I'd done on my final practice exam. And not just the math—I was pretty sure I'd scored far more points on the literature section than I had on the previous test. So I was in high spirits as I headed to the library after my classes.

"There you are!" Tetra greeted me as I walked in. Her smile was almost painfully bright.

I walked over and sat next to her. "How are you always so chipper?"

"Am I? I don't know, I just have a happy personality, I guess." Her smile broadened even more.

"I wish I could say the same. Anyway, what are you working on today?"

"Something called the 'probability density function of a normal distribution,' apparently." She handed me a card.

Probability density function of a normal distribution

The probability density function $f(x)$ of a normal distribution with mean μ and standard deviation σ is

$$f(x) = \frac{1}{\sqrt{2\pi\sigma^2}} \exp\left(-\frac{(x-\mu)^2}{2\sigma^2}\right).$$

"You're studying statistics now?" I asked.

"Not really. I told Mr. Muraki I wanted something that looked hard, something with lots of letters in it, and he gave me this card."

"Why'd you ask for something like that?"

"Well, you know how I tend to get all lost when I'm faced with a bunch of variables? I want to do something about that, so I figured with some practice..."

"I see. It's definitely important to keep your cool and look closely at the form of things when faced with a difficult problem," I said, recalling my test from the day before. "The form of an equation can give you clues as to where you need to go. Like, even if you don't know what a probability density function is, you can use its form to learn something about this function $f(x)$. First off, it's important to note where on the right side the x of $f(x)$ appears."

"That would be... here!" Tetra pointed at the card.

$$f(x) = \frac{1}{\sqrt{2\pi\sigma^2}} \exp\left(-\frac{(x-\mu)^2}{2\sigma^2}\right)$$

"Well spotted. This is important because it lets us think about how different values of x will affect the function overall."

"Yep, I was thinking the same thing. For example, this function has an $x - \mu$ in it, so this part will equal 0 when $x = \mu$."

$$f(x) = \frac{1}{\sqrt{2\pi\sigma^2}} \exp\left(-\frac{(x-\mu)^2}{2\sigma^2}\right)$$

"That's right. What else?"

"Well, if we move out a bit, there's an $(x - \mu)^2$."

$$f(x) = \frac{1}{\sqrt{2\pi\sigma^2}} \exp\left(-\frac{(x-\mu)^2}{2\sigma^2}\right)$$

"And this part will be at least zero for any real value of x, because it's a square, right?" Tetra said.

I nodded silently.

"So this part in parentheses, the exp part, will always be 0 or less."

$$f(x) = \frac{1}{\sqrt{2\pi\sigma^2}} \exp\left(-\frac{(x-\mu)^2}{2\sigma^2}\right)$$

"It will," I said. "You've also found some symmetry."

"Symmetry how?"

"Symmetry because $(x-\mu)^2$ is in the form of a square. That means in the $y = f(x)$ graph there will be left–right symmetry around an $x = \mu$ axis."

"Oh, of course. Anyway, I've got more."

"Sure, sorry. Go ahead."

"Okay, so we can rewrite the exponential part like this."

$$-\frac{(x-\mu)^2}{2\sigma^2} = -\left(\frac{x-\mu}{\sqrt{2\sigma^2}}\right)^2$$

"Now I want to make a definition like this."

$$\begin{cases} \heartsuit = \dfrac{x-\mu}{\sqrt{2\sigma^2}} \\ \clubsuit = \dfrac{1}{\sqrt{2\pi\sigma^2}} \end{cases}$$

"Then the overall function $f(x)$ look like this."

$$f(x) = \clubsuit \exp\left(-\heartsuit^2\right)$$

"In other words..."

$$f(x) = \clubsuit\, e^{-\heartsuit^2}$$

"Okay, now I feel *so* much more comfortable," Tetra said. "I mean, look how nice and simple everything became!"

"Interesting. I like this a lot. This also lets us see the function's asymptotic behavior. When $x \to \pm\infty$ we get $-\heartsuit^2 \to -\infty$, so $e^{-\heartsuit^2} \to 0$. Also, we get $e^{-\heartsuit^2} > 0$ for any real-number value of x, so we know the graph of $y = f(x)$ will asymptotically approach the x-axis. In fact, that's exactly what the probability density function of a normal distribution looks like—it's symmetric about $x = \mu$ and asymptotically approaches the x-axis."

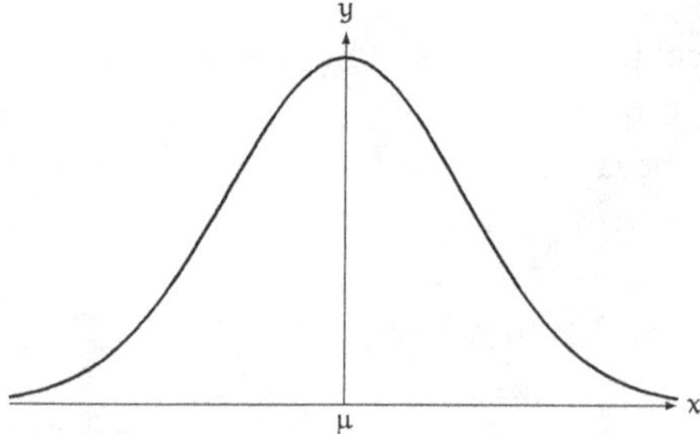

Graph of $y = f(x)$ for the probability density function of a normal distribution.

"No fair!" Tetra complained. "You're jumping ahead!"

"Oh, sorry. Once I start seeing the shape of a graph I can't help myself. I mean, it's all right there, right? Where the function increases and decreases, the symmetry, asymptotic curves—"

"Anyway, like I was saying, I was thinking about how when we define chunks of math as symbols like ♡ and ♣, it makes it a lot easier to see the form of the whole thing. Of course, it still feels strange to me that bringing new characters into this mess makes it easier to understand..."

"I think that's because you were the one who came up with the definitions, rather than just using a definition someone else made for you. It was you who saw the form of what's going on here—that's why you were able to make those definitions in the first place. In other words, introducing these new characters is proof you've discovered the overall structure here."

"Oh, I hadn't thought of it that way. But sure, I see that!"

"Another one of those discoveries you love so much," I said, smiling.

"I do indeed... Oh, but there's one part I'm still not totally sure about, here where I defined ♣."

$$\frac{1}{\sqrt{2\pi\sigma^2}}$$

"What's does this $\sqrt{2\pi\sigma^2}$ in the denominator mean?" she asked.

"Ah, that comes from the fact that $f(x)$ is a probability density function, which maps the set of all real numbers to the set of all nonnegative real numbers, and the value of its integral from $-\infty$ to ∞ equals 1."

"I'm . . . not really sure what that means."

"Well, we're kind of back to talking about definitions, but generally speaking a probability density function shows the probability $\Pr(\alpha \leqslant x \leqslant \beta)$ that a random variable x will take a value such that $\alpha \leqslant x \leqslant \beta$. We can represent that like this."

$$\Pr(\alpha \leqslant x \leqslant \beta) = \int_{\alpha}^{\beta} f(x)\,dx$$

"When we graph a probability density function, the area under the curve for $\alpha \leqslant x \leqslant \beta$ equals the probability."

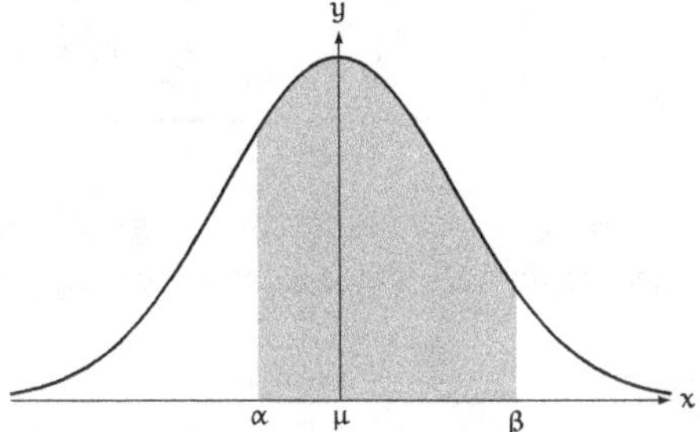

The area of a probability density function represents the probability $\Pr(\alpha \leqslant x \leqslant \beta)$.

"Okay . . ." Tetra said.

"So this means the value of the integral of a probability density function from $-\infty$ to ∞ equals 1, which is the probability that x will take *some* value. In the case of a normal distribution, this will always be true."

$$\frac{1}{\sqrt{2\pi\sigma^2}} \int_{-\infty}^{\infty} \exp\left(-\frac{(x-\mu)^2}{2\sigma^2}\right) dx = 1$$

"So that mysterious $\sqrt{2\pi\sigma^2}$ that's bothering you is just the value that makes $f(x)$ a probability distribution function.

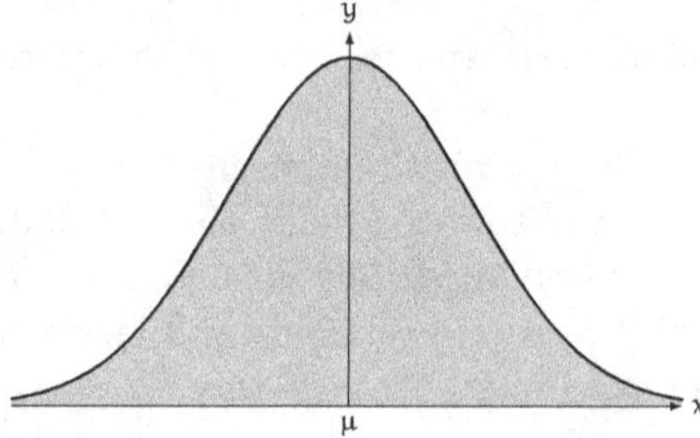

The value of the integral of a probability density function from $-\infty$ to ∞ equals 1.

"Huh," Tetra said. "So I guess we can define the definite integral part here as ♠."

$$\frac{1}{\sqrt{2\pi\sigma^2}} \underbrace{\int_{-\infty}^{\infty} \exp\left(-\frac{(x-\mu)^2}{2\sigma^2}\right) dx}_{\spadesuit} = 1$$

"Then the value of ♠ will be $\sqrt{2\pi\sigma^2}$, right?" She opened her notebook to a new page, and wrote a large new equation.

$$\int_{-\infty}^{\infty} \exp\left(-\frac{(x-\mu)^2}{2\sigma^2}\right) dx = \sqrt{2\pi\sigma^2} \qquad \cdots (♪)$$

"Sure, you can do that," I said. "And this should hold true for any values of μ and $\sigma \neq 0$. So long as $f(x)$ is a probability density function, the definite integral must have this value."

"Did you get your A?" came a voice from behind, startling me.

9.3.2 Reading the Laplace Integral

I spun around in my seat and saw Miruka standing there.

"I just took the test yesterday," I said. "They don't grade them *that* fast."

"Hmph." Miruka took a seat across from Tetra and me.

"I think I did well, though," I said. "There was a problem involving a definite integral with parameters, but even that didn't faze me."

"Definite integral with parameters, huh? That reminds me of the Laplace integral."

Miruka commandeered our notebook and wrote a definition.

The Laplace integral

For a real number a,

$$\int_0^\infty e^{-x^2}\cos 2ax\,dx = \frac{\sqrt{\pi}}{2}e^{-a^2}.$$

"You find this using integration by parts?" I asked.

"Indeed, you can. But only after *differentiating* by a. Let's start by letting $I(a)$ be the definite integral."

$$I(a) = \int_0^\infty e^{-x^2}\cos 2ax\,dx \qquad \cdots(\star)$$

"Now, differentiate $I(a)$ with respect to a."

$$\begin{aligned}
\frac{d}{da}I(a) &= \frac{d}{da}\left(\int_0^\infty e^{-x^2}\cos 2ax\,dx\right) && \text{differentiate}\\
&= \int_0^\infty \frac{\partial}{\partial a}\left(e^{-x^2}\cos 2ax\right)dx && \text{swap derivative and integral}\\
&= \int_0^\infty e^{-x^2}\frac{\partial}{\partial a}(\cos 2ax)\,dx && \text{here, } e^{-x^2} \text{ is a constant}\\
&= \int_0^\infty e^{-x^2}(-2x\sin 2ax)\,dx && \text{differentiate } \cos 2ax
\end{aligned}$$

"Technically we need some theory to justify that swap of the derivative and the integral, but let's save that for another day. Also, we're using the partial derivative $\frac{\partial}{\partial a}$ here, because what we're differentiating is a bivariate function in a and x. After integration by parts, the $I(a)$ shows up."

$$
\begin{aligned}
\frac{d}{da}I(a) &= \int_0^\infty e^{-x^2}(-2x\sin 2ax)\,dx && \text{from above}\\
&= \int_0^\infty (-2xe^{-x^2})\sin 2ax\,dx\\
&= \int_0^\infty (e^{-x^2})'\sin 2ax\,dx && \text{b/c } -2xe^{-x^2} = (e^{-x^2})'\\
&= \Big[e^{-x^2}\sin 2ax\Big]_0^\infty - \int_0^\infty e^{-x^2}(\sin 2ax)'\,dx && \text{integration by parts}\\
&= \Big[e^{-x^2}\sin 2ax\Big]_0^\infty - 2a\int_0^\infty e^{-x^2}\cos 2ax\,dx && \text{b/c } (\sin 2ax)' = 2a\cos 2ax\\
&= -2a\underbrace{\int_0^\infty e^{-x^2}\cos 2ax\,dx}_{I(a)}\\
&= -2aI(a)
\end{aligned}
$$

"So after differentiation with respect to a, then integration by parts with respect to x, we get this."

$$\frac{d}{da}I(a) = -2aI(a)$$

"We can think of this as a differential equation that considers $I(a)$ as a function of a. So let's solve it. If we let

$$y = I(a),$$

the differential equation takes the form

$$\frac{dy}{da} = -2ay.$$

Let's assume $y > 0$, and perform integration by substitution."

$$\begin{aligned}
\frac{dy}{da} &= -2ay \\
\frac{1}{y}\frac{dy}{da} &= -2a \\
\int \frac{1}{y}\frac{dy}{da}\,da &= \int -2a\,da && \text{integrate} \\
\int \frac{1}{y}\,dy &= -2\int a\,da && \text{integration by substitution} \\
\log y &= -a^2 + C_1 && C_1 \text{ is a constant} \\
y &= e^{-a^2+C_1} \\
&= e^{-a^2}e^{C_1} \\
&= Ce^{-a^2} && \text{let } C = e^{C_1}
\end{aligned}$$

"Since $y = I(a)$, we get

$$I(a) = Ce^{-a^2}.$$

So what should we do with this constant C, Tetra?"

"Maybe let $a = 0$?" Tetra replied. "Then we get $C = I(0)$."

$$\begin{aligned}
I(a) &= Ce^{-a^2} && \text{from above} \\
I(0) &= Ce^{-0^2} && \text{let } a = 0 \\
&= C
\end{aligned}$$

"Very good. In other words, we can see that the $I(a)$ we're after looks like this."

$$I(a) = I(0)e^{-a^2}$$

Miruka stopped talking and looked at me, then Tetra.

"Hang on a sec," I said. "Can't we be a little more concrete here? I mean, $I(a)$ started out as a definite integral."

$$\begin{aligned}
I(a) &= \int_0^\infty e^{-x^2}\cos 2ax\,dx && \text{from} \star \text{ on p. 291} \\
I(0) &= \int_0^\infty e^{-x^2}\,dx && \text{because } \cos 2ax = 1 \text{ when } a = 0
\end{aligned}$$

"So $I(0)$ should have the concrete value—"

"Stop right there," Miruka said. "Let Tetra tell us what that value would be."

"Who me?" Tetra said. "You want me to calculate *this*?"

$$I(0) = \int_0^\infty e^{-x^2}\,dx$$

Miruka and I remained silent, watching Tetra. She looked down at the notebook and nodded deeply. Flipping back a page, she pointed at something she'd written there.

"This one here, right? I think we can get $I(0)$ from this!"

$$\int_{-\infty}^{\infty} \exp\left(-\frac{(x-\mu)^2}{2\sigma^2}\right) dx = \sqrt{2\pi\sigma^2}$$ ♪ on p. 290

"This should hold for $\mu = 0$ and $\sigma = \frac{1}{\sqrt{2}}$. In this case $2\sigma^2 = 1$, so..."

$$\int_{-\infty}^{\infty} \exp\left(-x^2\right) dx = \sqrt{\pi} \qquad \text{letting } \mu = 0, \sigma = \frac{1}{\sqrt{2}}$$

"In other words, we get this integral over $-\infty$ to ∞."

$$\int_{-\infty}^{\infty} e^{-x^2}\,dx = \sqrt{\pi}$$

"From symmetry, the value over 0 to ∞ will be half that."

$$\int_0^\infty e^{-x^2}\,dx = \frac{\sqrt{\pi}}{2}$$

"So here's the value of $I(0)$!"

$$I(0) = \int_0^\infty e^{-x^2}\,dx = \frac{\sqrt{\pi}}{2}$$

"Well done," Miruka said. "Now we can use $I(0)$ to find $I(a)$. That's the Laplace integral."

$$I(a) = \int_0^\infty e^{-x^2} \cos 2ax\,dx = I(0)e^{-a^2} = \frac{\sqrt{\pi}}{2}e^{-a^2}$$

The Laplace integral (again)

For a real number a,

$$\int_0^\infty e^{-x^2} \cos 2ax \, dx = \frac{\sqrt{\pi}}{2} e^{-a^2}.$$

"Along the way, we also found something called the Gaussian integral."

The Gaussian integral

$$\int_{-\infty}^\infty e^{-x^2} \, dx = \sqrt{\pi}$$

"In this case we found the value of the Gaussian integral knowing that $f(x)$ is the probability density function of a normal distribution, but normally it goes the other way around—you find the value of the Gaussian integral separately, then use that to show that $f(x)$ is a probability density function."

"Laplace, Gaussian... So many kinds of integrals," Tetra said.

"Each one representing an enormous amount of work," Miruka added. "Enough to have one's name attached to it."

"And become a part of history," Tetra said with a sigh. "A history that so many mathematicians built up, piece by piece."

9.4 Fourier Series

9.4.1 Inspiration

"I had a problem involving definite integrals on my test yesterday," I said. "It took a little perspiration to solve it, but nothing crazy hard like all this."

"The problem actually made you sweat?" Tetra asked.

"Metaphorically speaking, I mean. The tedium of making your way through a long series of calculations without making any dumb mistakes. It didn't really take any inspiration, in other words. It looked something like this."

$$I(a, b) = \int_{-\pi}^{\pi} \left(a + b\cos x - x^2\right)^2 \, dx$$

"The a and b are real-number parameters, and I had to find the minimum value of $I(a, b)$ and the values of a and b at that minimum."[1]

Miruka looked at the problem I'd written with a gleam in her eye. "Hmmm..."

"Wow," Tetra said. "You can remember a problem this complex well enough to just write it out like that?"

"Well, I did it just yesterday, and spent a bit of time focused on it, so I guess it stuck with me. How would you tackle a problem like that?"

"I guess I would expand the $\left(a + b\cos x - x^2\right)^2$ and just work through the integrals of each term that popped out."

"Sure, that's what I did too. The expansion of $I(a, b)$ gives a quadratic in a and b, so all you have to do is complete the square. I think the answer was—"

"$a = \frac{\pi^2}{3}$," Miruka muttered.

"Huh, what was that?"

"$I(a, b)$ takes its minimum value when $a = \frac{\pi^2}{3}$, right?"

"What the...? How did you...?"

"From the b, of course," she said, touching a finger to her lips.

"From... Huh? I don't..." I was in something of a state of shock. I'm sure the calculations I'd gone through were nothing for her, but she hadn't written a thing! Surely even the mighty Miruka couldn't have worked all that out in her head, from expansion to integration to completing the square...

"Did you just do that in your head?!" an equally surprised Tetra said, echoing my thoughts.

"More like I peeked into the problem-writer's mind. I think this problem was written with the Fourier series expansion of x^2 in mind."

"Fourier series?" Tetra and I both said.

---

[1]See p. 280.

9.4.2 *Fourier Series*

"Before we talk about Fourier series," Miruka said, "Tetra, do you remember when we talked about Taylor series?"[2]

"Of course! I'll never forget it!" she replied. "You use Taylor series to turn functions like $\sin x$ into power series, right?"

The Taylor series (Maclaurin series) for $\sin x$

$$\sin x = +\frac{x}{1!} - \frac{x^3}{3!} + \frac{x^5}{5!} - \frac{x^7}{7!} + \cdots$$

"Generally, the Taylor series for a function $f(x)$ is this," Miruka said.

$$f(x) = \sum_{k=0}^{\infty} \frac{f^{(k)}(a)}{k!}(x-a)^k$$

"When $a = 0$, this is called a Maclaurin series. So what Tetra wrote here is the Maclaurin series for the function $\sin x$."

The Maclaurin series (Taylor series with $a = 0$) for $f(x)$

$$\begin{aligned} f(x) &= \frac{f(0)}{0!}x^0 + \frac{f'(0)}{1!}x^1 + \frac{f''(0)}{2!}x^2 + \cdots + \frac{f^{(k)}(0)}{k!}x^k + \cdots \\ &= \sum_{k=0}^{\infty} \frac{f^{(k)}(0)}{k!}x^k \\ &= \sum_{k=0}^{\infty} c_k x^k \qquad \text{where } c_k = \frac{f^{(k)}(0)}{k!} \end{aligned}$$

"Right," Tetra said. "Because $f^{(k)}(0)$ is the kth derivative of $f(x)$ when $x = 0$."

---

[2]See *Math Girls*, Ch. 9.

"This is how a Taylor series represents a function $f(x)$ as a power series. In contrast, a Fourier series represents $f(x)$ as a series of trigonometric functions."

The Fourier series for $f(x)$

$$\begin{aligned} f(x) &= (a_0 \cos 0x + b_0 \sin 0x) \\ &\quad + (a_1 \cos 1x + b_1 \sin 1x) \\ &\quad + (a_2 \cos 2x + b_2 \sin 2x) \\ &\quad + \cdots + (a_k \cos kx + b_k \sin kx) + \cdots \\ &= \sum_{k=0}^{\infty} (a_k \cos kx + b_k \sin kx) \end{aligned}$$

I looked back and forth, comparing the definitions for Taylor series and Fourier series.

$$f(x) = \sum_{k=0}^{\infty} c_k x^k \qquad \text{Taylor (Maclaurin) series for } f(x)$$

$$f(x) = \sum_{k=0}^{\infty} (a_k \cos kx + b_k \sin kx) \qquad \text{Fourier series for } f(x)$$

Okay, sure. Power series in x *above, trigonometric functions below... Got it.*

"So many letters..." Tetra said. "I'm getting dizzy!"

"Tetra..." I whispered.

"Oh, right. I'm supposed to be getting over that. Okay, so what kind of numbers are the a_k and b_k in the Fourier series?"

"You tell me," Miruka said kindly.

"I was afraid you'd say that... Right, let me think about this." After a moment, she said, "So the c_k in the Taylor series is a number we created from $f(x)$. So I wonder if maybe the a_k and b_k in the Fourier series are also numbers we're getting from $f(x)$?"

"Exactly," Miruka said. "Except that this time, we're *integrating*, not differentiating. So the a_k and b_k in the Fourier series are numbers created from integrals of $f(x)$. They're called Fourier coefficients."

Fourier coefficients and Fourier series

Letting $f(x)$ be a function that is expandable as a Fourier series, define sequences $\langle a_n \rangle$ and $\langle b_n \rangle$ (Fourier coefficients) as

$$\begin{cases} a_0 = \dfrac{1}{2\pi}\displaystyle\int_{-\pi}^{\pi} f(x)\,dx \\ b_0 = 0 \\ a_n = \dfrac{1}{\pi}\displaystyle\int_{-\pi}^{\pi} f(x)\cos nx\,dx \\ b_n = \dfrac{1}{\pi}\displaystyle\int_{-\pi}^{\pi} f(x)\sin nx\,dx \qquad (n = 1, 2, 3, \ldots) \end{cases}$$

Then,

$$f(x) = \sum_{k=0}^{\infty}(a_k \cos kx + b_k \sin kx)$$

(a Fourier series).

"So Taylor series use differentiation, and Fourier series use integration," I said.

"We find Fourier coefficients by multiplying $f(x)$ by $\cos nx$ and $\sin nx$, then integrating over $-\pi$ to π," Miruka said. "For example, let's consider the case where n is a positive integer, and we multiply by $\sin nx$."

$$f(x) = \sum_{k=0}^{\infty}(a_k \cos kx + b_k \sin kx)$$

"When we multiply both sides by $\sin nx$ and integrate, we get this."

$$\int_{-\pi}^{\pi} f(x)\sin nx\,dx = \int_{-\pi}^{\pi}\left(\sum_{k=0}^{\infty}(a_k \cos kx \sin nx + b_k \sin kx \sin nx)\right)dx$$

"Next, we swap the integral and the infinite series. There are some conditions on being able to do this, but trust me that they're satisfied here."

$$= \sum_{k=0}^{\infty} \left(\int_{-\pi}^{\pi} (a_k \cos kx \sin nx + b_k \sin kx \sin nx)\, dx \right)$$

"Let's separate the integral into (cs) and (ss), and see what values they take."

$$= \sum_{k=0}^{\infty} \left(a_k \underbrace{\int_{-\pi}^{\pi} \cos kx \sin nx\, dx}_{\text{(cs)}} + b_k \underbrace{\int_{-\pi}^{\pi} \sin kx \sin nx\, dx}_{\text{(ss)}} \right)$$

"This k moves in the range of nonnegative integers, but interestingly enough (cs) becomes 0 and disappears. Also, (ss) remains as π only when $k = n$, but otherwise it too becomes 0 and disappears."

$$= b_n \int_{-\pi}^{\pi} \sin nx \sin nx\, dx$$

$$= b_n \pi$$

"In other words, everything other than the $\sin nx \sin nx$ integral just goes poof."

"Huh. I remember working out that integral on my test yesterday ..."[3]

"We can use everything up to this point to find b_n."

$$b_n = \frac{1}{\pi} \int_{-\pi}^{\pi} f(x) \sin nx\, dx \qquad (n = 1, 2, 3, \ldots)$$

"And similarly to find a_n ... "

$$a_n = \frac{1}{\pi} \int_{-\pi}^{\pi} f(x) \cos nx\, dx \qquad (n = 1, 2, 3, \ldots)$$

"A rather circuitous route, but now we have a_n and b_n for integer n."

"I guess a_0 gets special treatment?" Tetra asked.

[3]See pg. 279.

"It does. For a_0, we use the fact that $\cos 0x \cos 0x$ equals 1."

$$
\begin{aligned}
\int_{-\pi}^{\pi} f(x)\,dx &= \int_{-\pi}^{\pi} f(x)\cos 0x\,dx \\
&= \underset{\sim\sim\sim}{a_0 \int_{-\pi}^{\pi} \cos 0x \cos 0x\,dx} \quad \text{only } k = 0 \text{ term remains} \\
&= a_0 \int_{-\pi}^{\pi} 1\,dx \\
&= a_0 \Big[x \Big]_{-\pi}^{\pi} \\
&= a_0 (\pi - (-\pi)) \\
&= a_0 \cdot 2\pi \\
a_0 &= \frac{1}{2\pi} \int_{-\pi}^{\pi} f(x)\,dx
\end{aligned}
$$

"Too bad we can't also make a_0 and b_0 the same," Tetra said.

"If you want to unify Fourier coefficients, you can just give the $n = 0$ case some special treatment in the Fourier series."

Fourier coefficients and Fourier series (alternative representation)

Fourier coefficients

$$
\begin{cases}
a'_n = \dfrac{1}{\pi} \displaystyle\int_{-\pi}^{\pi} f(x)\cos nx\,dx \\
b'_n = \dfrac{1}{\pi} \displaystyle\int_{-\pi}^{\pi} f(x)\sin nx\,dx
\end{cases}
\qquad (n = 0, 1, 2, \ldots)
$$

Fourier series

$$
f(x) = \frac{a'_0}{2} + \sum_{k=1}^{\infty} (a'_k \cos kx + b'_k \sin kx)
$$

9.4.3 Beyond Perspiration

"Okay, Miruka," I said, "I get that Fourier series use trigonometric functions, but I'm still not seeing the relation between them and my test problem."

"Your problem looked like this, right?" Miruka said.

$$I(a, b) = \int_{-\pi}^{\pi} \left(a + b\cos x - x^2\right)^2 \, dx$$

"Let's let a be $a \cos 0x$, and rewrite a, b as a_0, a_1."

$$I(a_0, a_1) = \int_{-\pi}^{\pi} \left(a_0 \cos 0x + a_1 \cos 1x - x^2\right)^2 \, dx$$

"x^2 is an even function, so when we expand x^2 as a Fourier series we'll only see the even function $\cos kx$. Specifically, the Fourier series for x^2 is this."

$$x^2 = a_0 \cos 0x + a_1 \cos 1x + a_2 \cos 2x + \cdots$$

"See how the first two terms are the same?"

"Huh, they are..."

"Wait, they're the same how?" Tetra asked.

"Think about what this definite integral $I(a_0, a_1)$ gives us," Miruka said. "It finds the difference between $a_0 \cos 0x + a_1 \cos 1x$ and x^2, squares that, and returns the integral of the result. It's a kind of error evaluation. His problem said to find the values of a_0, a_1 that minimize the integral, so it seems reasonable to look at the a_0, a_1 Fourier coefficients."

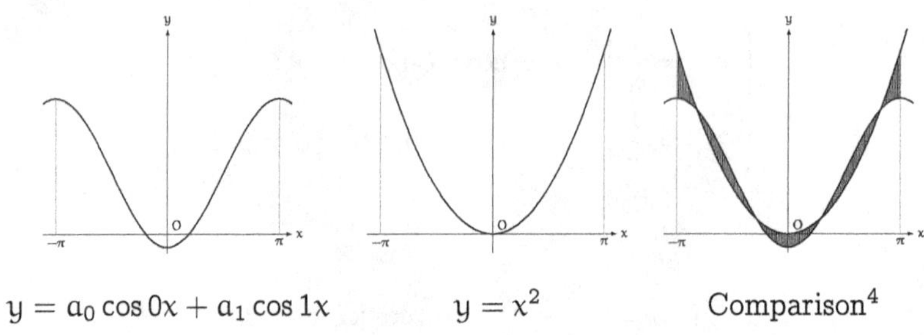

$y = a_0 \cos 0x + a_1 \cos 1x$ $\qquad$ $y = x^2$ $\qquad$ Comparison[4]

Here we go again. Miruka receiving inspiration from... somewhere.

[4]Note that we integrate after squaring, so this area itself is not the error.

"Correct me if I'm wrong," Tetra said, "but what we're saying here is if we know the Fourier series for x^2, the first two terms will be pretty close to x^2, right? But what I definitely don't get is how anyone could be expected to notice something like that while taking a test. I know I never would!"

"I don't think that was the intent on his test," Miruka said. "That problem probably really was just to see how well you can grind through a hairy integral. They didn't expect him to haul out Fourier series. I just wanted to show how seeing the form of the problem can make it more fun."

"But you still haven't told us how you were able to just whip out that $a = \frac{\pi^2}{3}$," I said. "Did you seriously take the integral for the a_0 Fourier coefficient in your head?"

$$a_0 = \frac{1}{2\pi} \int_{-\pi}^{\pi} f(x) \cos 0x \, dx$$

Miruka smiled. "That's not too hard for $f(x) = x^2$ and $\cos 0x = 1$, is it?"

$$a_0 = \frac{1}{2\pi} \int_{-\pi}^{\pi} x^2 \, dx$$

"Oh... right. I guess that's not so bad after all..."

$$\begin{aligned} a_0 &= \frac{1}{2\pi} \int_{-\pi}^{\pi} x^2 \, dx \\ &= \frac{1}{2\pi} \cdot \frac{1}{3} \Big[x^3 \Big]_{-\pi}^{\pi} \\ &= \frac{1}{6\pi} \left(\pi^3 - (-\pi)^3 \right) \\ &= \frac{\pi^2}{3} \end{aligned}$$

9.4.4 Beyond Inspiration

"Maybe I'm just stating the obvious," Tetra said, "but it seems like a single mathematical statement is often saying multiple things, not just one."

"Well of course," Miruka said. "A mathematical statement can represent infinitely many things, both what its author intended and otherwise. In some cases, it takes centuries to discover meanings even the author wasn't aware of."

"Right! We were just recently talking about prophetic discoveries!"

"It would be foolish to limit ourselves to one interpretation of what we read. Not that we can read anything we want into mathematical statements, but there's often plenty we can logically deduce."

"I see... By the way, one thing that was bothering me—it seems like x^2 is pretty simple as it is. Why would we ever want to go to the trouble of representing it as trigonometric functions? I understand Taylor series, why we might want to represent something hard like $\sin x$ as something simpler like x^k, but the other way around...?"

Miruka grinned and pushed her glasses back with a finger. "Well, for example we can use the Fourier series for x^2 if we want to make you stand up from your chair, speechless."

Tetra held up a hand in protest. "Oh, come on. I'm not *that* excitable... usually."

"We'll see..." Miruka took a minute to perform some calculations on a nearby piece of paper, then looked back up. "Okay. So here's what you get after expanding x^2 into a Fourier series."

$$\begin{aligned} x^2 &= \frac{\pi^2}{3} + 4\left(\frac{-\cos 1x}{1^2} + \frac{+\cos 2x}{2^2} + \frac{-\cos 3x}{3^3} + \cdots + \frac{(-1)^k \cos kx}{k^2} + \cdots\right) \\ &= \frac{\pi^2}{3} + 4\sum_{k=1}^{\infty} \frac{(-1)^k \cos kx}{k^2} \end{aligned}$$

"Let's say $x = \pi$. Noting that $\cos k\pi = (-1)^k$, we get this."

$$\begin{aligned} x^2 &= \frac{\pi^2}{3} + 4\sum_{k=1}^{\infty} \frac{(-1)^k \cos kx}{k^2} && \text{from above} \\ \pi^2 &= \frac{\pi^2}{3} + 4\sum_{k=1}^{\infty} \frac{(-1)^k(-1)^k}{k^2} && \text{let } x = \pi \\ &= \frac{\pi^2}{3} + 4\sum_{k=1}^{\infty} \frac{(-1)^{2k}}{k^2} \\ &= \frac{\pi^2}{3} + 4\sum_{k=1}^{\infty} \frac{1}{k^2} && \text{because } (-1)^{2k} = 1 \end{aligned}$$

"From that, we can derive this."

$$\sum_{k=1}^{\infty} \frac{1}{k^2} = \frac{1}{4}\left(\pi^2 - \frac{\pi^2}{3}\right) = \frac{\pi^2}{6}$$

"Look familiar, Tetra?"

$$\sum_{k=1}^{\infty} \frac{1}{k^2} = \frac{\pi^2}{6}$$

Tetra sprang from her chair, gasping and making sounds that weren't quite words.

"Hey, th-that's..." I stammered. "That's the Basel problem!"[5]

"Indeed it is. The very problem that stumped mathematicians for over a hundred years, until Euler finally provided a proof in the eighteenth century. So from the Fourier series for x^2 we got this form of the answer to the Basel problem."

$$\sum_{k=1}^{\infty} \frac{1}{k^2} = \frac{\pi^2}{6}$$

"We can also write it as a zeta function, like this."

$$\zeta(2) = \frac{\pi^2}{6}$$

Tetra was still standing, silently staring with wide-open eyes at what Miruka had written.

"Told ya," Miruka said.

"So the Fourier series for x^2 gives the value of $\zeta(2)$..." I said.

When we'd started talking about this, I certainly hadn't expected to bump into Euler and his zeta functions. Not in the slightest.

Powers of x, trigonometric functions, differentiation, integration—the way that mathematical concepts are so subtly intertwined never failed to amaze me. Problems solved by inspiration, those requiring perspiration, both were equally astonishing. But even all that was just a tiny part of mathematics. The world of mathematics was something beyond inspiration and perspiration. Something far wider and deeper than humanity itself.

[5] See *Math Girls*, Ch. 9.

A great discovery solves a great problem, but there is a grain of discovery in the solution of any problem.

George Pólya
How to Solve It

CHAPTER **10**

The Poincaré Conjecture

> Thus, the implementation of Hamilton's program would imply the geometrization conjecture for closed three-manifolds.
>
> GRIGORI PERELMAN
> *The entropy formula for the Ricci flow and its geometric applications*
> http://arxiv.org/abs/math/0211159

10.1 THE OPEN SEMINAR

10.1.1 After the Lecture

The usual gang and I attended an open seminar, hosted annually at a local university. The speakers were faculty at the university, but their presentations were tailored to a general audience. This was our second time attending; the previous year, the theme had been Fermat's last theorem, and this year it was the Poincaré conjecture.

Wow, it's been a year already?

After the first presentation, which had lasted a little over an hour, we filed out of the auditorium, along with other small groups of high-school students and others who had come to listen.

When we made it outside, Yuri stretched her arms wide. "I have no idea what I just heard," she said, her breath steaming in the frosty air.

"Yeah, the video was interesting," Tetra said, "but I'm not sure I really understood what it was trying to say. Sure is cold out here!" She pulled out a stick of lip balm and started applying it.

"Lunchtime," Lisa said, hoisting her red backpack.

"I heard that!" Yuri said. "Let's go see what's on the menu."

"Hmph," Miruka muttered.

The five of us cut across the campus, heading for the school's dining hall.

10.1.2 Lunchtime

"Seems like just yesterday we were here hearing about Fermat's last theorem," I said between mouthfuls of pilaf.

"It does, doesn't it," Yuri said over her dish of spaghetti carbonara. "Is it just me, or was that one a lot easier to follow? With this Poincaré stuff, I don't even understand what the problem is!"

"I thought last year's problem was hard because it involved so many equations," Tetra said, poking at her rice omelette. "This year there were nearly no equations, but that certainly didn't make it any easier to understand! That film they showed, with the rocket in space. Was that supposed to be, like, actual outer space? The one we're in? After that was something about heat and temperature and how they're described in physics, but I got totally lost trying to figure out how that tied in with the math."

"Metaphor?" Lisa said from behind her sandwich.

"I also got lost with all the names that came up," Yuri said. "I figured out that it was Poincaré's problem but somebody named Perelman came up with the proof, but all those others? No clue."

Miruka sat quietly through our chatter. After taking her time finishing off her slice of chocolate cake, she closed her eyes. We all fell silent, as if she had sent us a signal.

"What is form?" she finally said.

And with that, the space around us transformed into our own private lecture hall.

10.2 Poincaré

10.2.1 Form

"What is form?" Miruka repeated. "Not an easy question to answer. Form is so pervasive around us that it's difficult to describe it.

"Being asked to describe form is a lot like being asked to describe numbers. You might try to answer by giving some specific numbers, like $1, 2, 3$, but a handful of examples like that doesn't really answer the question. Providing some examples of numbers is important, sure, but that alone isn't enough to think deeply about the issue.

"In algebra, we consider constructs like groups and rings and fields. We think about operations between elements, we establish axioms regarding their use, and we consider what we can say as a result. This allows us to decompose the many characteristics condensed into the word 'number' into smaller parts, which we can use to rebuild that concept piece-by-piece. Through study like that, we can close in on just what numbers are. It even leads us to discoveries surpassing the concepts of 'number' we are familiar with.

"So I ask you once again—what is form? There are so many concepts behind that word. Length, size, angle, orientation, front, back, surface area, volume, congruence, similarity, bending, twisting, joining, affixing, separating, connecting, cutting... I could go on. To answer my question, we need to decompose these many characteristics condensed into the word 'form' into smaller parts, then rebuild that concept piece-by-piece. Through study like that, we can close in on just what form is. It may even lead us to discoveries surpassing the concepts of 'form' we are familiar with."

Miruka paused, allowing Tetra to ask, "So, if we're studying form—we're talking about geometry here?"

"In a sense," Miruka said. "Today's seminar was all about topology, which is an offshoot of geometry."

"But Miruka, math is math, right?" Yuri said. "What's this about studying all these separate concepts?"

Miruka turned to face her. "Math is math, but math is also broad. It contains many areas to explore. When you approach form from different directions, you end up focusing on different properties. When we equip a set with a topology to define a topological space, we can think about continuity and connectivity. When we use that topological space to define a manifold, we can think about dimensions. If we go on to define differentiable manifolds, we can think about things like differentiation and tangent spaces. We can add Riemannian metrics to define Riemannian manifolds, allowing us to think about distances and angles and curvature. There are many objects we can study through geometry, but each and every one is in some sense a representation of form. And sometimes we want to focus on just one particular characteristic, setting others aside."

"Like the Bridges of Königsberg problem?" Yuri asked. "Because we could move things around, so long as we don't change how everything's connected."

"Truly an apropos example," Miruka said with a soft smile. "The Bridges of Königsberg problem was the genesis of graph theory, which in turn is considered the origin of topology. We have Euler to thank for that. But anyway, you're correct that the problem considers all forms with the same connections as being the same. In other words, so long as you don't change any connections, you can change shapes however you like. In that case we're talking about graph isomorphism, but we do something similar in topology—we're interested in topological invariants, which are quantities that remain unchanged under deformation of a space into some other homeomorphic space."

"And things that don't change are worthy of naming," I added.

"Once we've settled on what things we're going to consider as being the same," Miruka continued, "it's natural to want to start putting things into categories. We start gathering things up and classifying them all, something like classical natural history."

"Like collecting and classifying butterflies!" Tetra said.

"Or insects," I said.

"Or gemstones," Yuri said.

"Or gears," Lisa said.

"That's right," Miruka said. "The study of form starts with putting each in a specimen box labeled with its name. Categorization is thus the first step toward the study of form. Classification of two-dimensional closed surfaces finished in the nineteenth century. This classification was according to orientability and genus, namely whether the surface has front and back sides and the number of holes it contains. Of course, many mathematicians were involved in this effort. They also knew that closed two-dimensional manifolds could also be classified from another perspective—which of three possible geometric structures it contained."

Miruka paused to scan our faces before continuing.

"The Poincaré conjecture is a fundamental problem related to the classification of three-dimensional manifolds. A fundamental problem, yes, and a natural question to ask, but one that's easily answered? No, not at all. In fact, it would take a century of work by many mathematicians to finally answer it."

10.2.2 The Poincaré Conjecture

"Before we get to the answer, could you please explain what the problem was?" Yuri said. "I'm still stuck on that!"

"I think that was somewhere in the handouts we got at the seminar, right?" Tetra said, digging through her huge pink bag. "Handouts are definitely important for this kind of thing, aren't they," she muttered. "Now where did I put that... ?"

Lisa slid her copy of the handout across the table.

"That's the one! Thanks!" Tetra said. "Let's see... So here it says the Poincaré conjecture comes from a paper written by a mathematician named Henri Poincaré. That paper was written right at the beginning of the twentieth century, in 1904, and is considered a very important paper in the field of topology. The problem itself is... oh, right here!"

The Poincaré conjecture

If the fundamental group for a closed 3-manifold M is isomorphic to the trivial group, then M is homeomorphic to the 3-sphere.

"Oookaaay..." Yuri said. "So, like I said, can someone please explain the problem?"

"Well, I think I can at least explain some of the terms," Tetra said. "We can think of a closed 3-manifold as a space that's bounded and has no boundary, where locally it looks like you're in a three-dimensional Euclidean space. And fundamental groups are those groups we made based on loops, right?"

"It's kind of hard to imagine a closed 3-manifold," I said, hoping to help out, "but it's like two filled globes side-by-side with their surfaces attached. Bounded, but with no boundaries."

Yuri just groaned.

"Understanding what the terms mean is important," Miruka said, "but let's start by getting a grip on the Poincaré conjecture's logical structure. Maybe it's easier to see what's going on if we rewrite it like this."

The Poincaré conjecture (rephrased)

Assume conditions $\mathsf{P}(\mathsf{M})$ and $\mathsf{Q}(\mathsf{M})$ for a closed 3-manifold M:

$\mathsf{P}(\mathsf{M}) =$ The fundamental group for M is isomorphic to the trivial group.

$\mathsf{Q}(\mathsf{M}) = \mathsf{M}$ is homeomorphic to the 3-sphere

Then,

$$\mathsf{P}(\mathsf{M}) \Longrightarrow \mathsf{Q}(\mathsf{M})$$

holds for M.

"That's right!" Tetra said. "And we can say the reverse from the fact that the fundamental group is a topological invariant, right?"[1]

"From that *with* the fact that the fundamental group for a 3-sphere is isomorphic to the trivial group," Miruka said.

The fundamental group is a topological invariant

Because the fundamental group for a 3-sphere is isomorphic to the trivial group, and the fundamental group is a topological invariant,

$$\mathsf{P}(\mathsf{M}) \Longleftarrow \mathsf{Q}(\mathsf{M}).$$

[1]See p. 207.

"I'm just getting more and more lost!" Yuri cried.

"There's nothing hard here," Miruka said. "We're just talking about logical structure, which you're good at. We just want to say these two things."

- $\mathsf{P}(M) \Longrightarrow \mathsf{Q}(M)$ (what the Poincaré conjecture says)
- $\mathsf{P}(M) \Longleftarrow \mathsf{Q}(M)$ (what we know is true)

"So what could we say if we were to prove the Poincaré conjecture to be true?"

"Uh, that $\mathsf{P}(M)$ and $\mathsf{Q}(M)$ are the same?"

"That's right. Proving the Poincaré conjecture to be true would assure us that $\mathsf{P}(M)$ and $\mathsf{Q}(M)$ are essentially the same conditions. We would be able to state this."

$$\mathsf{P}(M) \Longleftrightarrow \mathsf{Q}(M)$$

The fundamental group for M is isomorphic to the trivial group	$\Longleftrightarrow$	M is homeomorphic to the 3-sphere

"Sure, I understand that, but... so?"

Unable to contain myself, I jumped in. "The Poincaré conjecture is an attempt at better understanding the power of this tool called the fundamental group. If it holds, then if you want to investigate whether M is homeomorphic to the 3-sphere, all you have to do is check if its fundamental group is the trivial group."

"In his attempts to build up the field of topology," Miruka said, "Poincaré used a tool called homology groups as a way of classifying and identifying manifolds. He was able to use homology groups to classify *two*-dimensional closed manifolds, but not three-dimensional ones. In fact, Poincaré himself discovered a dodecahedral space that served as a counterexample demonstrating that such a thing was impossible.

"So, he turned to fundamental groups. He asked, might it be possible to use fundamental groups to classify *all* closed 3-manifolds? This was a significant problem. In his paper, Poincaré posed this as the problem of whether we can use fundamental groups to determine if there is a homeomorphism with the 3-sphere."

"Is that different from categorizing?" Yuri asked.

"It is. If the statements 'the fundamental groups of M and N are isomorphic' and 'M and N are homeomorphic' both have the same truth value, that's a categorization by fundamental groups. In contrast, if the statements 'the fundamental group of M is isomorphic to the trivial group' and 'M is homeomorphic to the 3-sphere' have the same truth value, that's using fundamental groups to determine if there is a homeomorphism with the 3-sphere. See the difference, Yuri?"

"I guess classification is... harder?"

"Sure. If we can categorize by fundamental groups, we can use them to determine homeomorphism with the 3-sphere. However, it's been proven that closed 3-manifolds cannot be categorized using fundamental groups, taking as a counterexample lens spaces, a discovery that followed Poincaré's death. Okay, so if we can't use fundamental groups to *categorize* closed 3-manifolds, maybe we can at least use them to *determine* homeomorphism with the simplest 3-manifold, the 3-sphere. That's what the Poincaré conjecture asks."

"Maybe we should back up to what a fundamental group is," Yuri said.

"Like Tetra said, it's a group we create from loops," I said. "We consider all loops that can be continuously deformed into each other as being the same, and take the operation of joining them together as the group operation. So a closed 3-manifold with the trivial group as its fundamental group would be a space in which we could continuously deform every possible loop down to a single point. Say you launch a rocket with a really long string attached to it, and send it off into space. When it travels for a while and returns, you reel the string back in, making the loop smaller and smaller. Would the string get stuck on anything? What $P(M)$ says is that no matter what path the rocket takes through M, you'll always be able to reel the loop back in without it getting caught. If the Poincaré conjecture holds, we can say that always being able to reel the loop in would mean M is homeomorphic to the 3-sphere, while getting caught up somewhere would mean it isn't."

"Okay, I think I get that," Yuri said. "So that guy Perelman, he proved that we can use fundamental groups for this identification?"

"He did, but that's not all," Miruka said. "What he actually proved was Thurston's geometrization conjecture, a more generalized statement that includes the Poincaré conjecture. So by proving the one, he proved the other."

"Thurston?" Tetra said.

"Yet another name," groaned Yuri.

10.2.3 Thurston's Geometrization Conjecture

"Thurston's geometrization conjecture is an even broader statement than the Poincaré conjecture," Miruka said.

Thurston's geometrization conjecture

All closed 3-manifolds have a prime decomposition into pieces having one of eight geometric structures.

"Say you have a closed 3-manifold M, and you want to know whether it's homeomorphic to the 3-sphere," Miruka said. "The Poincaré conjecture asks whether we can use the fundamental group to make that determination. In contrast, Thurston's geometrization conjecture is about classification of closed 3-manifolds. It asks whether every 3-manifold has a prime decomposition into pieces having one of eight geometric structures."

"Oh, that sounds like integer decomposition!" Tetra said.

"What do you mean?" Yuri asked.

"You know, how we can factor integers into primes! Doesn't Thurston's geometrization conjecture sound like a similar kind of thing?"

"Sure, it's similar in a sense," Miruka said. "A closed 3-manifold can be uniquely decomposed into a connected sum of prime manifolds. By a connected sum, I mean removing a ball from each of two manifolds and joining them at the resulting boundaries. It's a method of continuing decomposition down to a prime decomposition. If Thurston's geometrization conjecture holds, when we reach that point, each piece will have one of eight geometric structures. We can also look at it as a conjecture that, just as we can use primes to characterize every integer, we can use these eight geometric structures to characterize every 3-manifold. Further, by proving Thurston's geometrization conjecture, one would also prove the Poincaré conjecture."

"What determines a geometric structure?" Tetra asked.

"What congruence means in that space. We can use a congruence transformation group to represent that using groups. A geometric structure is the combination of a space along with its congruence transformation group. Geometric structures arose from the Erlangen program, an effort by Klein to characterize geometries according to invariants in their transformation groups. You have to be careful, because 'geometry' can have various meanings. For example, in the early twentieth century we learned that closed 2-manifolds have three kinds of geometric structures—spherical, Euclidean, and hyperbolic. The congruence transformation group for spherical geometry is $SO(3)$, in Euclidean geometry it's the Euclidean congruence transformation group, and in hyperbolic geometry it's the hyperbolic transformation group $SL_2(\mathbb{R})$. You can think of this as a way of using groups to classify geometries, and you can think of Thurston's geometrization conjecture as a version of that for closed 3-manifolds. With the exception that you're breaking things down into pieces, of course."

"Like taking apart a clock," Lisa muttered.

10.2.4 Hamilton's Ricci Flow

"Let me make sure I've got all this straight," I said. "The Poincaré conjecture is about whether we can use fundamental groups to determine if closed 3-manifolds are homeomorphic with the 3-sphere. Thurston's geometrization conjecture is about whether closed 3-manifolds can be decomposed into pieces having one of eight types. Perelman proved the latter, which simultaneously proved the former."

"You've got it. But before we get to Perelman, we need to talk about Hamilton."

"*Another* name?" Yuri said.

"Oh, there was something about Hamilton in the handout," Tetra said, skimming through the pamphlet. "It says he used something called Ricci flow in his attempt to prove Thurston's geometrization conjecture, and produced several results along that line. He was able to prove the Poincaré conjecture with the condition of something called positive Ricci curvature. In other cases there were some outstanding questions, though, so the problem remained unresolved for twenty years. Perelman was the one who finally filled in those last remaining holes, creating a new method to finally prove Thurston's geometrization conjecture."

"What a convoluted mess!" Yuri said.

"Okay, my turn to make sure I've got everything straight," Tetra said, jotting down some notes. "It goes like this, right?"

- Poincaré came up with the Poincaré conjecture, but he wasn't able to prove it himself.
- Thurston came up with Thurston's geometrization conjecture, but he wasn't able to prove it himself, either.
- Hamilton used something called Ricci flow to prove the Poincaré conjecture with some conditions, but he never found a way to get rid of those conditions.
- Perelman used Hamilton's Ricci flow to prove Thurston's geometrization conjecture, along with the Poincaré conjecture with no conditions.

"Now *this* I can understand," Yuri said.

"It's kind of like passing the baton in a relay race!" Tetra said. "Just because you come up with a good problem, that doesn't mean you'll be able to solve it on your own. Sometimes you have to just pass it on to others. It's like all of these mathematicians are working together, passing the baton from one to another."

"It is," Lisa said.

"Not quite," Yuri objected. "I'm pretty sure each would have rather crossed the goal line with that baton still in hand."

"Even so, Tetra's summary is more or less correct," Miruka said with a serious expression. "My only objection would be that it implies it was just these four mathematicians who did the whole thing. It did take many years for the Poincaré conjecture to be proved in three dimensions, but research by many mathematicians gave proofs in higher dimensions. Many mathematicians also performed detailed investigations of the eight geometric structures of Thurston's geometrization conjecture, confirming many cases where it does hold. These of course included Thurston himself. So it's not like everyone was sitting around, waiting for a proof to materialize."

"I still think it's all a big mess," Yuri said.

"Sometimes history is messy," Miruka said, "even when we'd prefer something simpler."

10.3 Mathematicians

10.3.1 A Chronology

Miruka turned to Lisa. "Mind pulling up that chronology?"

"Done," Lisa said, turning her computer so we could see its screen.

We all leaned in to get a better look.

Year	Event
~300 BC	Euclid's *Elements* is compiled.
1736	Euler's paper describing the Seven Bridges of Königsberg problem is published.
18th cent.	Many mathematicians, including Saccheri, Lambert, Legendre, Farkas Bolyai (father of János), d'Alembert, and Thibaut, attempt to prove the parallel postulate, but none are successful.
1807	Fourier applies Fourier series to the heat equation.
1813	Gauss apparently discovers non-Euclidean geometries, but does not publish his results.
1822	Fourier's *The Analytical Theory of Heat* is published.
1824	Bolyai discovers non-Euclidean geometries.
1829	Lobachevsky publishes a paper on non-Euclidean geometries.

~1830	Galois theory is developed.
1832	Bolyai publishes his work on non-Euclidean geometries.
1854	Riemann describes manifolds in his inaugural lecture.
1858	Listing and Möbius independently discover the Möbius strip.
1861	Listing describes the Möbius strip in a published paper.
1865	Möbius describes the Möbius strip in a published paper.
1860s	Möbius classifies closed 2-manifolds by genus.
1872	Klein advocates the Erlangen program in his inaugural lecture.
1895	Poincaré writes the first paper on topology.
19th cent.	Klein, Poincaré, Beltrami, and others construct models of non-Euclidean geometries.
1904	Poincaré writes his fifth paper, describing Poincaré dodecahedral space and the Poincaré conjecture.
1907	Poincaré, Klein, and Koebe classify closed 2-manifolds according to three geometric structures (Euclidean, spherical, and hyperbolic geometries).
1961	Smale publishes a proof of the Poincaré conjecture for five and higher dimensions.
1966	Smale is awarded the Fields Medal.
1980	Thurston proposes his geometrization conjecture.
1980	Hamilton introduces Ricci flow.
1980	Hamilton proves the Poincaré conjecture with the condition of positive Ricci curvature.
1982	Freedman proves the Poincaré conjecture in four dimensions.
1982	Thurston publishes a paper regarding his geometrization conjecture.
1982	Thurston is awarded the Fields Medal.

1990s	Hamilton applies Ricci flow to closed 2-manifolds.
2000	The Clay Mathematics Institute publishes its Millennium Prize Problems, which include the Poincaré conjecture.
2002	Perelman writes a paper announcing the Hamilton program.
2003	Perelman publishes two more papers.
2006	The International Congress of Mathematicians confirms Perelman's proof.
2006	Perelman is awarded, but refuses, the Fields Medal.
2007	Morgan and Tian publish an analysis of Perelman's proof.
2010	The Clay Mathematics Institute announces Perelman as winner of the first Millennium Prize for his resolution of the Poincaré conjecture.
2010	Perelman refuses the Millennium Prize.

"Of course, this is just a teeny tiny bit of the full history of this problem," Miruka said. "I haven't even touched on all those who took on the Poincaré conjecture, but failed to prove it."

"It says this Mr. Smale solved the Poincaré conjecture for five and higher dimensions," Tetra said. "Does that mean there are different kinds of Poincaré conjecture?"

"The Poincaré conjecture is an assertion about closed 3-manifolds, but you can generalize by changing the 3 to n. Of course, when doing so you also have to generalize the fundamental group."

"Also, Mr. Freedman gave his proof for four dimensions *after* Mr. Smale's proof for five or more dimensions. It seems kind of strange that the lower-dimension proofs come after the higher-dimension ones."

"And the final proof for three dimensions, the problem as it was originally stated, came only after all that," Miruka said. "Funny how that works sometimes, isn't it."

10.3.2 The Fields Medal

"Wow, things really started moving in the 1980s, didn't they," I said. "Thurston's geometrization conjecture comes out, then Hamilton's Ricci flow..."

"What's the Fields Medal?" Yuri asked.

"That's the one that's sometimes called the Nobel prize for mathematics, right?" Tetra said. "It's the most famous prize in the field of mathematics."

"There's an age restriction on it, though," I said. "You have to receive it by age forty. I read that Wiles was just past forty when he proved Fermat's last theorem, so he got a special commendation in its place."

"Wait, it says Perelman was refused the Fields Medal. Because he was past forty too?"

"He wasn't refused it, he refused it himself despite being offered the award in 2006."

"What the...? Why on earth would he do such a thing?"

"We can't really be sure," Miruka said. "Some say he thought Hamilton wasn't being sufficiently credited for his contributions, others say it's because he was uncomfortable with all the hoopla not directly related to mathematics."

"The Fields Medal website," Lisa said, turning her computer to show us.

We looked at the section for 2006, which showed Perelman's name along with the other three award recipients for that year, listed in alphabetical order.

2006

Andrei Okounkov

Grigori Perelman*

Terence Tao

Wendelin Werner

*Grigori Perelman declined to accept the Fields Medal.

Above that was an image of the award itself.

The Fields Medal, depicting a profile of Archimedes. (Image by Stefan Zachow, ZIB)

10.3.3 The Millennium Prize

"So what's this one?" Yuri asked, pointing to the mention of the Millennium Prize in the chronology.

"It's a cash prize you can receive by solving one of seven unsolved problems from a list the Clay Mathematics Institute published in 2000," I said. "The Poincaré conjecture was on that list, but again Perelman refused the award."

"A cash prize, huh? How much?"

"One million dollars. So the CMI has a reserve of seven million dollars to cover the entire list."

"A *million dollars* for solving one math problem?! Okay, I'm definitely going to solve *at least* one."

"There are some rules about who gets the money," I said, looking at a website Lisa had already pulled up. "First, your solution to the problem has to be published in a refereed journal. That's to ensure that specialists in the field have confirmed your solution, I suppose. You also have to wait two years after being published to ensure the mathematics community has accepted your solution. After that, the CMI consults with field experts to determine whether the award should be given."

"But still... *a million dollars*!"

"There's a curious relationship between mathematicians and awards," Miruka said. "We want to reward people who have accomplished great things, and there are many awards given for solutions to mathematical problems. Big cash prizes like the Millennium Prize are intriguing, but nobody becomes a mathematician for the money. We don't do research for financial gain. We do math because it's wonderful, because it's something we love."

"Well, not to argue or anything," Tetra said, "but I think the relationship between humanity and mathematics is a bit more complex than that. Even a mathematician working alone on a problem isn't really working alone. I think this chronology makes that very clear. It's more like countless mathematicians working in cooperation, at a level beyond space and time."

"Hmph."

"If Perelman really did refuse his awards because he didn't think Hamilton was getting enough credit, isn't that a good example of how important he considers the contributions of other mathematicians?"

"It is."

"I think that attitude shows a lot of respect for the work of others, as well as a love of mathematics. The same holds for not just solving a problem, but writing up your solution and sharing it with the world. We've talked several times about how mathematics transcends time, but I think that's specifically because of this cooperation by so many mathematicians."

"You're exactly right," Miruka said. "And Perelman's paper does include a section where he carefully traces through the work by his predecessors. Everyone contributes differently. What's important is whether your contribution enriches the world of mathematics. Fermat's last theorem produced many mathematicians, and the Bridges of Königsberg led to the field of topology, but these examples show it's difficult to predict which mathematics will lead to enrichment. In the end, mathematics is a tightly woven tapestry. Some will add a thread here or there, others will lay out a design. Everyone is contributing to its completion, according to their interests and abilities."

"But when everyone just follows their own interests, doesn't that lead to fields of mathematics becoming increasingly niche and specialized?"

"That's one way to look at it. Another is to say that they're deepening their areas in preparation for solving important problems later on."

"Algebraic topology is the application of methods from algebra to topology, right?" I asked, changing the subject.

"Oh, so algebra and geometry cooperated to prove the Poincaré conjecture?!" Tetra said.

"That's half right, half less so" Miruka said, making a chopping gesture. "It did require many mathematicians advancing different fields in mathematics to solve the Poincaré conjecture. Algebraic topology, differential topology, and others still. But despite all that, the method used to finally prove the Poincaré conjecture came from a different field entirely—physics."

Miruka stood and brushed her hair back in an ebon wave. We each looked up and followed her with our eyes.

"We've talked many times about how mathematicians build bridges between worlds. Sometimes solving a problem in one area of mathematics requires techniques from another. Nothing wrong with that. It's a good thing, even. Connecting different areas makes the overall stronger."

"We use whatever weapons are at hand!" Tetra said.

"Kaboom," Yuri added.

10.4 Hamilton

10.4.1 Ricci Flow

I realized how long we had been sitting and talking in the university cafeteria. My drink had long gone cold. I was too entranced to get up and replace it, though.

"Grigori Perelman's achievement was to close up some final gaps that required filling to prove Thurston's geometrization conjecture and the Poincaré conjecture. Before we talk about that, however, we need to talk about Richard Hamilton, because he's the one who discovered the tool that was absolutely necessary to complete that proof. A tool called Ricci flow."

Miruka paused to take a sip of her coffee. Tetra took advantage of that pause to slip in a question.

"I remember they mentioned that in the seminar. That's what you mean by a method from physics, right? I didn't understand a word they said about it. I mean, I'm sure it's a difficult thing to understand in the first place, but even harder to understand is how you can use physics in a math proof. I've always thought of mathematics as something that isn't governed by physical laws."

Miruka nodded in acceptance of Tetra's concerns. "It's not that physical laws were used for a mathematical proof. Hamilton's Ricci flow has a form similar to Fourier's heat equation, which was discovered in physics through the study of heat transfer. That similarity is in the form of the differential equations, and the behavior of the functions that solve them, but it isn't like we're treating the Poincaré conjecture as a physics problem."

"Fourier's heat equation, huh?" Tetra said.

"Oh yay," Yuri muttered. "Another name."

10.4.2 Fourier's Heat Equation

"Physics is highly concerned with physical quantities," Miruka quietly said. "Time, position, speed, acceleration, pressure, temperature, things like that. A physics problem generally starts with considering something like a position x of some object at time t and temperature u. The physical laws governing such situations are often described in the form of differential equations."

We silently nodded.

"When we use Fourier's heat equation, we consider an object's temperature as a function of position and time, and we express that function as a differential equation. Given an initial temperature distribution for some object, we can imagine how its temperature will change over time."

"Like a hot cup of coffee gradually cooling?" Tetra asked.

"Following Newton's law of cooling?" I asked.

"Except that Newton's law of cooling only considers time," Miruka said. "Fourier's heat equation also considers position."

"Like when the coffee's hot, but the cup is cold?" Yuri asked.

"Or in heat conduction experiments, where you heat one end of a metal rod and investigate how the heat conducts!" I said. "The rate of conduction depends on what material the rod is made from."

Miruka nodded. "A good example."

"Oh, I think I'm getting this!" Tetra said, her face brightening. "Newton's equations of motion and Hooke's law represent positions as differential equations. Fourier's heat equation is doing the same thing, but for temperature instead!"

10.4.3 A Reversal of Ideas

Tetra's happy expression soon clouded over.

"Er, on second thought, maybe I'm not getting this quite as well as I'd thought. I get that Fourier's heat equation uses differential equations for temperature. But even if Hamilton's Ricci flow is similar to that, surely it doesn't use differential equations to describe temperature, right? I mean, a purely mathematical object can't have a temperature, so..."

"Think of it more like an analogy—temperature is to the heat equation what a Riemannian metric is to the Ricci flow equation." Miruka said. "A Riemannian metric—devised by Riemann, as you might expect—is information describing distance and curvature on a manifold."

"One more new name for the list..." Yuri said.

Tetra groaned and held her head. "But something still feels weird. Temperature changing over time is a real-world physical law, so that's fine. But isn't it strange to say that this Riemannian metric—not that I really understand what one is, but some mathematical object—changes over time? It still feels like we're saying the world of mathematics is governed by physical laws!"

Tetra's words reverberated within me, particularly because they felt so true. I didn't really understand these concepts either, but I could sympathize with how she was trying to maintain consistency in her understanding. Even if we can substitute temperature with a Riemannian metric, she was asking, can we really consider a mathematical object as changing over time? Once again, she had impressed me.

"Your doubts are certainly justified," Miruka said, a gleam in her eyes. "It is something of a reversal of ideas. The heat equation is a differential equation related to temperature, and we can investigate temperature changes by examining it. But the Ricci flow is a little different. The objective of a Ricci flow equation is to change the Riemannian metric. The goal isn't to discover change, it's to induce change."

"I still don't understand," Tetra said. "I mean, time is—"

"Time t in a Ricci flow equation isn't physical time, it's just a parameter, a convenient metaphor, allowing use of expressions like 'ancient solutions' and 'initial conditions' that are applied to this concept. But we are *not* using actual time in mathematical proofs or anything."

"I see," Tetra said with a small nod.

"So in a Ricci flow equation, we consider a Riemannian metric having t as a parameter. This Riemannian metric determines curvature, so if it changes, curvature does too. When curvature changes, so does the manifold. What we want to do using Ricci flow is to skillfully change the Riemannian metric—in other words, to skillfully change the curvature, and thereby skillfully change the manifold."

"Skillfully, in what sense?" Tetra immediately asked.

"In the sense that it allows us to prove Thurston's geometrization conjecture. There are infinitely many ways in which we can change a Riemannian metric. Hamilton described the direction of change in a Ricci flow. A Riemannian metric having a parameter t and satisfying a Ricci flow equation is called a Ricci flow. Hamilton hoped that a Ricci flow could make the curvature of a manifold uniform, kind of how the passage of time makes temperature within an object uniform. It's much easier to do math on a manifold with uniform curvature. Indeed, manifolds with constant curvature have been classified since the mid-twentieth century."

"I'm still just relieved to hear that t isn't real time," Tetra said. "Actually, I just remembered how we used a parameter called t when we were talking about loops. That wasn't real time either, but I guess we can kinda think of it that way, can't we?"

"The t in loops doesn't have anything to do with the t in Ricci flow, but they're the same in the sense that they're both parameters. Anyway, Hamilton discovered the Ricci flow equation, studied it, and tried to find a way to apply it to a proof of Thurston's geometrization conjecture. That attempt was called Hamilton's program."

10.4.4 Hamilton's Program

Hamilton's program

By changing the Riemannian metric via the Ricci flow equation, we will deform a closed 3-manifold and thereby derive a solution to the Thurston geometrization conjecture. We will remove any singularities arising during deformation through a process called *surgery*.

"Hamilton used the parameter t in the Ricci flow equation to change the Riemannian metric. Doing so allows control of the curvature in a closed 3-manifold. There are various kinds of curvature. Riemann's curvature tensor contains all the information about the bending of a Riemannian manifold, but it is a delicate and complicated thing, making it difficult to handle. What Hamilton did is condense a curvature tensor R_{ijkl}, allowing its manipulation as a quantity called the Ricci curvature R_{ij}. The hope was that in the solution to a Ricci flow equation, the Ricci curvature would gradually even out over time, eventually becoming uniform. One problem, however, was that when doing so, singularities—sites of infinite curvature—might arise. To address this, Hamilton devised something called 'Ricci flow with surgery.' In this method, we stop time just before a singularity is about to occur, perform surgery to excise it, then resume time."

"By time, you mean this parameter t, right?" Tetra asked.

"That's right. The correspondence goes something like this." Miruka began writing.

The world of physics		The world of mathematics
Heat equation	←----→	Ricci flow equation
Heat conductor	←----→	Closed 3-manifold
Position x	←----→	Position x
Time t	←----→	Parameter t
Temperature	←----→	Riemannian metric (for deriving Ricci curvature)
Averaged temperature	←----→	Averaged Ricci curvature

"So, what are we doing here?" I said. "I'm not seeing the relationship between making curvatures uniform and proving Thurston's geometrization conjecture."

"The hope here is that regardless of what kind of Riemannian metric a closed 3-manifold has, applying a Ricci flow to deform it will lead to deriving a closed 3-manifold with uniform curvature. This would be effective for organizing and classifying the infinitely many closed 3-manifolds."

"But is it really that easy, Miruka?" Yuri asked. "All it takes is this surgery thing?"

"Hamilton's work produced many proofs. First off, he showed that there exists a Ricci flow with a given Riemannian metric as its initial value. He also proved that the Poincaré conjecture holds under the condition that the Ricci curvature is positive everywhere. Furthermore, he proved that Thurston's geometrization conjecture holds if there are no singularities and the sectional curvature is uniformly bounded."

"But still, only under certain conditions," Tetra said.

"That's right. Hamilton showed that both the Poincaré conjecture and Thurston's geometrization conjecture hold *under certain conditions*. But singularities still prevented completion of Hamilton's program, so Hamilton came up with the method of Ricci flow with surgery to remove them. What he needed to do was show that we can remove all singularities by a finite number of operations, and that the sectional curvatures remaining after that was done would be uniformly bounded. What he found, however, was that a kind of singularity called 'the cigar' could arise, and there was no way to handle those. So Hamilton's program was stalled, and it remained that way for two decades."

"Twenty years!" Yuri said.

"So Perelman found a way to remove the cigar singularities?" I asked.

"Even better. He showed that they never arise in the first place."

"Never?"

"Never. Hamilton hoped that was the case, but it took Perelman's 'no local collapsing' theorem to prove it. But that still wasn't enough. Perelman also proved his canonical neighborhood theorem and others along the way, but he finally completed Hamilton's program, proving Thurston's geometrization conjecture. Perelman published three papers related to Ricci flow. These papers proved Thurston's geometrization conjecture, proved the Poincaré conjecture, and gave power to Ricci flow."

"So this Mr. Perelman . . . he closed those final gaps?"

"I suppose that's one way of putting it," Miruka said, her voice lowering. "But I think one has to be quite deep in this field to be able to appraise just how big those gaps were. Whether we should say Perelman proved Thurston's geometrization conjecture, or that Hamilton and Perelman deserve equal credit, is a tricky question to answer. Without Hamilton's Ricci flow, Perelman wouldn't have found his clues, and without the methods and theorems that Perelman introduced the proof wouldn't have been completed. When such a significant proof has been made, how are we to compare the work by the person who closed the final gaps, and those who made all the discoveries leading up to those gaps? Of course, in mathematics the general custom is to give credit to whoever managed to take that final step—whether they want it or not."

"These papers?" Lisa asked with a slight cough. She was pointing at her computer screen.

Miruka glanced at the screen and nodded. "What say we take a look at them?"

10.5 Perelman

10.5.1 *Perelman's Papers*

We brought our heads together and looked at Lisa's computer.

- Perelman, Grisha. "The entropy formula for the Ricci flow and its geometric applications."[2]
- Perelman, Grisha. "Ricci flow with surgery on three-manifolds."[3]
- Perelman, Grisha. "Finite extinction time for the solutions to the Ricci flow on certain three-manifolds."[4]

"Grisha Perelman is a Russian mathematician. His first name is actually Grigory, but he published these under his nickname."

"These papers are in English?" Yuri asked.

"Well of course," I said. "They're certainly not in Japanese."

"Totally *not* what I meant," Yuri sniped. "Since he's Russian I figured he might write in his own language."

"Most papers today are written in English," Miruka said, "the lingua franca that allows people around the world to read them."

"So they become letters to humanity!" Tetra said, nodding.

"Perelman added some footnotes in his papers, giving thanks to several organizations that provided him with opportunities to perform his research. Let's read one."

> I was partially supported by personal savings accumulated during my visits to the Courant Institute in the Fall of 1992, to the SUNY at Stony Brook in the Spring of 1993, and to the UC at Berkeley as a Miller Fellow in 1993–95. I'd like to thank everyone who worked to make those opportunities available to me.

"Here's how his first paper begins."

[2]https://arxiv.org/abs/math/0211159
[3]https://arxiv.org/abs/math/0303109
[4]https://arxiv.org/abs/math/0307245

> The Ricci flow equation, introduced by Richard Hamilton, is the evolution equation $\frac{d}{dt}g_{ij}(t) = -2R_{ij}$ for a Riemannian metric $g_{ij}(t)$.[5]

"Wow, Miruka, you're amazing!" Yuri squealed.

"Not at all. I'm just reading what someone else wrote. The content of this paper is, quite frankly, beyond my current skillset. Even so, a bit of browsing reveals parts we can understand. Like this."

> Thus, the implementation of Hamilton's program would imply the geometrization conjecture for closed three-manifolds. In this paper, we carry out some details of Hamilton's program.

"Here, Perelman is stating the procedure by which he will prove Thurston's geometrization conjecture, namely, by implementing Hamilton's program."

"Here he says 'we carry out,'" Tetra said. "Did he have a co-author?"

"That's something called the 'author's we' you'll sometimes see in academic writing," Miruka said. "It's used even when there's just a single author."

" 'We' to indicate yourself? Very strange..."

"Perelman was the only author, but he isn't the only reader. In this case, 'we' is meant to indicate both the author and the reader. It means those making their way through this paper are there alongside Perelman, working toward implementation of Hamilton's program."

"I like that!" Tetra said. "The reader not being alone, I mean, and cooperating with the author to work on a problem!"

"You can also think of it as the author asking the reader to participate in the discussion," Miruka said.

"Implement?" Lisa said, raising her voice to indicate a question.

"The Hamilton program points the way toward a proof of Thurston's geometrization conjecture. That alone isn't enough,

---

[5] What Perelman is writing as $\frac{d}{dt}$ here looks like ordinary differentiation, but because it includes partial differentiation within the Ricci curvature R_{ij}, it actually represents partial differentiation.

though—we need actual proofs. That's what he means by 'implement.'"

"You had Perelman's papers there on your computer, Lisa?" I asked.

"Just searched," she said, shrugging. "Easily found."

"Perelman wrote his proof in the form of these papers," Miruka said, "and posted them on the Internet to a website called arXiv. Anyone can download and read any paper posted to arXiv as a PDF, just like we've done here."

"Download? Yes. Read? I'm not so sure..." Yuri said.

"Once you have the English ability, you'll be able to read the words at least. And once you have the mathematical ability, you'll be able to follow Perelman's arguments. What's interesting in his case is that he just posted these papers to arXiv, not to a peer-reviewed journal."

"Hang on! Wasn't that one of the requirements for winning the million dollars? You said the proof has to be published in a journal."

"Not a problem. According to the rules for the Millennium Prize Problems, there are some alternatives. In this case, Morgan and Tian published an analysis of the problem, and that was taken as an acceptable substitute. Even so, Perelman refused both the Fields Medal and the Millennium Prize when they were offered to him.

"Huh, okay."

"I rather like the fact that Perelman posted his paper to arXiv. I think it's better for mathematics to post one's work in a form that's immediately available to mathematicians around the world. Perelman's papers included many examples of new techniques that had never been applied before. His use of this tool called Ricci flow expanded the world of mathematics, and mathematicians who applied Ricci flow in this new form expanded it even further."

10.5.2 One Step Further

"So as we've seen, Perelman resolved Thurston's geometrization conjecture and the Poincaré conjecture by posting his papers to arXiv," Miruka said. "But—"

We all leaned forward in anticipation of what would follow.

"But doesn't just knowing that leave you feeling a little... dissatisfied? Don't you want to try taking at least one step further into the math?"

"I sure do," Tetra said.

"You bet," I said.

"If it isn't *too* hard," Yuri said.

Lisa remained silent.

"Of course, working with Ricci flow means dealing with the partial differential equations of curvature tensors, heady stuff to just jump straight into. Beyond my understanding, honestly. So, what to do, what to do..."

"I know!" Tetra said, raising a hand. "I'd like to know more about methods from physics. We can find more living language there, I'd bet!"

"Sounds like a plan," Miruka immediately said.

"I'd like to get another drink first," Yuri said. "Math is thirsty work."

"We'll also need a lot more paper," Miruka said.

"Want me to look for some in the school store?" I asked.

"Here," Lisa said, pulling a sheaf of blank paper from her red backpack.

10.6 Fourier

10.6.1 Fourier's Era

"So did Fourier win the Fields Medal too?" Yuri asked after coming back with a glass of melon soda.

"No, no," I said. "He died long before that was a thing."

Lisa opened her computer and gave a slight cough. "Joseph Fourier," she read. "French mathematician and physicist. Born 1768, died 1830. John Fields, Canadian mathematician. Born 1863, died 1932. Fields Medal, established in 1936."

"So Mr. Fourier was from the late eighteenth to the early nineteenth century, right?" Tetra said. "Isn't that around the time of the French Revolution?"

"Fourier was born into a poor family, and orphaned at age eight," Miruka said. "I imagine he had a very rough life, but it was also a very eventful one. He became a professor of mathematics, participated in Napoleon's expedition to Egypt, and served as a prefectural governor. He was undoubtedly a man of many talents. I hear he even had a close encounter with the guillotine."

"The guillotine?!" Yuri shouted.

"In 1811 the Paris Academy of Sciences called for papers on heat conduction. Fourier submitted a paper on research he'd been conducting, and won."

I took a sip of my warm tea and thought of Fourier while looking out at the blue sky and soft winter sun. I wondered what his feelings about family might have been like, having been orphaned at eight years old. What kind of emotions did he feel when tackling mathematics? I found it difficult to imagine.

10.6.2 Fourier's Heat Equation

"Fourier's heat equation deals with temperature," Miruka said.

"We just recently talked about Newton's law of cooling," I said.

"Newton's law of cooling is represented as a function $u(t)$ for time t. In contrast, Fourier's heat equation involves a function $u(x, t)$ of position x and time t. Let's compare the two. First off, here's Newton's law of cooling, with room temperature set to 0."

Newton's law of cooling

The rate of temperature change is proportional to the difference in temperatures as

$$\frac{d}{dt}u(t) = Ku(t) \qquad \text{for a constant } K.$$

"The function $u(t)$ here represents the temperature of an object at time t. This differential equation represents the state of an object placed in a room at temperature 0."

"Got it."

"In contrast, here's Fourier's heat equation."

Fourier's heat equation

The temperature $u(x,t)$ satisfies the partial differential equation

$$\frac{\partial}{\partial t}u(x,t) = K\frac{\partial^2}{\partial x^2}u(x,t) \qquad \text{for a constant } K.$$

This is called Fourier's heat equation (in one dimension).

"So what exactly is this heat equation? You could say it's a partial differential equation representing the propagation of heat. Another way to describe it would be as a partial differential equation that *models* the physical phenomenon of heat propagation. So let's play a bit with Fourier's heat equation as an analogue of Ricci flow. Specifically, let's investigate this function $u(x,t)$.

"To set the stage, imagine you have an infinitely long, straight wire with some temperature distribution at time t. Those are the initial conditions. Since we have an initial temperature distribution, we know the starting temperature at any position x. As the time t changes, so will the temperature at positions along the wire. In other words, we're representing the temperature u at time t as a bivariate function $u(x,t)$. At time $t = 0$, there may be differences in temperature at different values for x—the wire may be hot in some locations, cold in others. As time t advances, however, we can predict that those differences will become smaller, eventually reaching a point where the temperature has evened out across the entire wire. What we want to do is take a closer look at the form of this function $u(x,t)$."

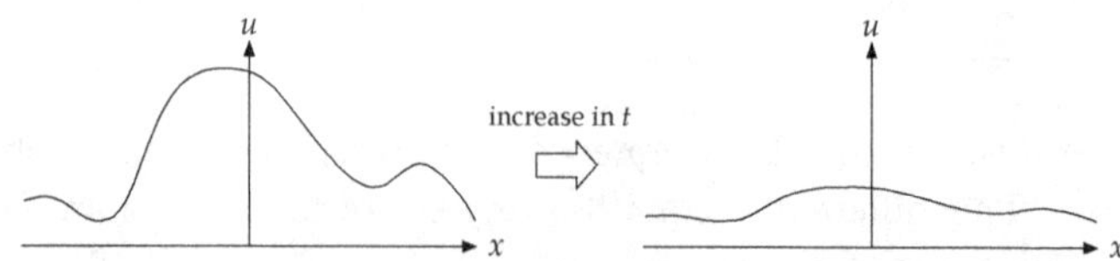

Temperature distribution in a wire and its change with increasing time t.

"This heat equation is also called the heat conduction equation or the heat diffusion equation. You can also use it for other things, like the way a scent will disperse in air. For simplicity, let's use $K = 1$ as the constant of proportionality."

$$\frac{\partial}{\partial t}u(x,t) = \frac{\partial^2}{\partial x^2}u(x,t) \qquad \text{heat equation with } K = 1$$

"To solve this equation, we need to find a function $u(x,t)$ that satisfies it. The $\frac{\partial}{\partial t}u(x,t)$ on the left shows a partial derivative of the function $u(x,t)$ with respect to t. The $\frac{\partial^2}{\partial x^2}u(x,t)$ on the right is the second partial derivative of $u(x,t)$ with respect to x."

"Hang on a sec," Tetra said, holding up a hand to stop Miruka. "Can we start with just what a *partial* derivative is?"

"Sure. This function $u(x,t)$ takes two variables. When we hold x constant and differentiate with respect to t, we call that the 'partial derivative of $u(x,t)$ with respect to t,' and write that as $\frac{\partial}{\partial t}u(x,t)$. Or, we can treat t as being constant and differentiate with respect to x, which would be the partial derivative $\frac{\partial}{\partial x}u(x,t)$. If we differentiate the result with respect to x once more, we write that as $\frac{\partial^2}{\partial x^2}u(x,t)$."

"It goes like this, right?" I said. "If we have a bivariate function like this..."

$$u(x,t) = x^3 + t^2 + 1$$

"...its partial derivatives would be this."

$$\begin{aligned}
u(x,t) &= x^3 + t^2 + 1 && \text{a bivariate function} \\
\frac{\partial}{\partial t}u(x,t) &= 2t && \text{partial derivative of } u(x,t) \text{ w.r.t. } t \\
\frac{\partial}{\partial x}u(x,t) &= 3x^2 && \text{partial derivative of } u(x,t) \text{ w.r.t. } x \\
\frac{\partial^2}{\partial x^2}u(x,t) &= 6x && \text{2nd partial derivative of } u(x,t) \text{ w.r.t. } x
\end{aligned}$$

Miruka nodded. "That's right. Partial differentiation with respect to t treats x as a constant, and partial differentiation with respect to x treats t as a constant. The superscripts show the number of times we're differentiating. For example, you could also write the heat equation like this."

$$\frac{\partial}{\partial t}u(x,t) = \frac{\partial^2}{\partial x^2}u(x,t)$$

"Or, we can use just omit the (x, t) part we already know about and just write the heat equation like this."

$$\frac{\partial u}{\partial t} = \frac{\partial^2 u}{\partial x^2}$$

"There are other ways we can write it as well."

"Okay, I'm good with that," Tetra said. "Sorry for interrupting your talk."

"I don't get *any* of this," Yuri said. "*Way* over my head."

"There are parts you can understand," Miruka said. "Let's just talk about the signs. If the $\frac{\partial u}{\partial t}$ on the left is positive, then the temperature at position x will increase as time t advances. If the $\frac{\partial^2 u}{\partial x^2}$ on the right is positive, that means the temperature at position x is lower than the mean temperature of the temperatures to the right and left of x. The heat equation binds the two together. That means if the temperature at a given position goes up at some point in the future, the temperature at that point must have been lower than the mean of the temperatures to the left and right. We can understand this as being a property of heat, and this partial derivative shows us in detail how that property behaves. Let's take one more look at the partial differential equation we're considering."

$$\frac{\partial}{\partial t}u(x, t) = \frac{\partial^2}{\partial x^2}u(x, t)$$

"Next, let's use separation of variables to find a solution, and use that to solve the heat equation."

10.6.3 Separation of Variables

"To use separation of variables, we assume this bivariate function $u(x, t)$ can be written as a product of two univariate functions in x and t, which we'll call $f(x)$ and $g(t)$."

$$u(x, t) = f(x)g(t)$$

"So let's rewrite the $u(x, t)$ in the heat equation as $f(x)g(t)$."

$$\begin{aligned} \frac{\partial}{\partial t}u(x, t) &= \frac{\partial^2}{\partial x^2}u(x, t) && \text{heat equation} \\ \frac{\partial}{\partial t}f(x)g(t) &= \frac{\partial^2}{\partial x^2}f(x)g(t) && \text{rewrite } u(x, t) \text{ as } f(x)g(t) \end{aligned}$$

"We can work out the left side. Since x is constant, we can treat $f(x)$ as a constant, too."

$$\begin{aligned}
\frac{\partial}{\partial t} f(x)g(t) &= f(x) \cdot \frac{\partial}{\partial t} g(t) && f(x) \text{ is constant} \\
&= f(x) \cdot \frac{d}{dt} g(t) && \text{ordinary derivative for univariate function } g \\
&= f(x)g'(t)
\end{aligned}$$

"On to the right side. This time t is constant, so we can treat $g(t)$ as constant, too."

$$\begin{aligned}
\frac{\partial^2}{\partial x^2} f(x)g(t) &= g(t) \cdot \frac{\partial^2}{\partial x^2} f(x) && g(t) \text{ is constant} \\
&= g(t) \cdot \frac{d^2}{dx^2} f(x) && \text{ordinary derivative for univariate function } f \\
&= f''(x)g(t)
\end{aligned}$$

"So here's what the heat equation has become."

$$f(x)g'(t) = f''(x)g(t)$$

"We'll assume $u \neq 0$, meaning $f(x) \neq 0$ and $g(t) \neq 0$. To separate out bits dependent on x and those dependent on t, we can divide both sides by $f(x)g(t)$ to get this."

$$\underbrace{\frac{g'(t)}{g(t)}}_{\text{depends only on } t} = \underbrace{\frac{f''(x)}{f(x)}}_{\text{depends only on } x}$$

"Take a close look at this equality. The left side depends only on t, meaning that if t doesn't change, its value will be constant no matter what happens to x. Similarly, the right side depends only on x, so if that value doesn't change the right side is constant regardless of changes in the t value. So in the end, we find that the value of this equality remains unchanged regardless of how x and t change. That's what's so nice about separation of variables. Let's write this constant as $-\omega^2$."

$$\frac{g'(t)}{g(t)} = \frac{f''(x)}{f(x)} = -\omega^2$$

"So now we've created two ordinary differential equations."

$$\begin{cases} f''(x) &= -\omega^2 f(x) \\ g'(t) &= -\omega^2 g(t) \end{cases}$$

"See what we've done? Through separation of variables, we changed one partial derivative of a bivariate function into two normal derivatives of univariate functions, derivatives we're very familiar with. We can use trigonometric functions for f and an exponential function for g to obtain general solutions for each."

$$\begin{cases} f(x) &= A\cos\omega x + B\sin\omega x \\ g(t) &= Ce^{-\omega^2 t} \end{cases}$$

"So, $u(x,t) = f(x)g(t)$ is the solution for the heat equation."

$$\begin{aligned} u(x,t) &= f(x)g(t) \\ &= (A\cos\omega x + B\sin\omega x)\cdot Ce^{-\omega^2 t} \\ &= e^{-\omega^2 t}(AC\cos\omega x + BC\sin\omega x) \\ &= e^{-\omega^2 t}(a\cos\omega x + b\sin\omega x) \qquad \text{let } a = AC, b = BC \end{aligned}$$

"There's our solution. Let's write this as $u_\omega(x,t)$."

$$u_\omega(x,t) = e^{-\omega^2 t}(a\cos\omega x + b\sin\omega x)$$

10.6.4 Superposition of Solutions by Integration

"This solution introduces parameters a, b, and ω," Miruka continued.

$$u_\omega(x,t) = e^{-\omega^2 t}(a\cos\omega x + b\sin\omega x)$$

"We want to consider a and b for each ω, so we'll write these as functions, $a(\omega)$ and $b(\omega)$."

$$u_\omega(x,t) = e^{-\omega^2 t}(a(\omega)\cos\omega x + b(\omega)\sin\omega x) \qquad \cdots ①$$

"This ω can take any nonnegative real value, and a superposition of the solutions as determined by each ω is also a solution. So we integrate by ω and use that superposition as the solution. We're doing this because we want to represent the initial conditions from this superposition."

$$\begin{aligned} u(x,t) &= \int_0^\infty u_\omega(x,t)\, d\omega && \text{superposition} \\ &= \int_0^\infty e^{-\omega^2 t}\left(a(\omega)\cos\omega x + b(\omega)\sin\omega x\right) d\omega && \text{from ①} \end{aligned}$$

"See how $e^{-\omega^2 t} = e^{-\omega^2 \cdot 0} = 1$ when $t = 0$? That means we can represent the temperature distribution when $t = 0$, our initial conditions, like this."

$$u(x,0) = \int_0^\infty \left(a(\omega)\cos\omega x + b(\omega)\sin\omega x\right) d\omega \qquad \text{initial condition}$$

"This is called the 'Fourier integral' representation."

10.6.5 The Fourier Integral

"This Fourier integral is a continuous version of the Fourier series. It's a form in which we've changed the sums in a Fourier series into an integral."

$$\begin{aligned} f(x) &= \sum_{k=0}^{\infty} \left(a_k \cos kx + b_k \sin kx\right) && \text{Fourier series for } f(x) \\ u(x,0) &= \int_0^\infty \left(a(\omega)\cos\omega x + b(\omega)\sin\omega x\right) d\omega && \text{Fourier integral for } u(x,0) \end{aligned}$$

"In a Fourier series, when we're given a function $f(x)$ we find the Fourier coefficients a_n and b_n by integration."

$$\begin{cases} a_0 = \dfrac{1}{2\pi}\displaystyle\int_{-\pi}^{\pi} f(x)\,dx \\ b_0 = 0 \\ a_n = \dfrac{1}{\pi}\displaystyle\int_{-\pi}^{\pi} f(x)\cos nx\,dx \\ b_n = \dfrac{1}{\pi}\displaystyle\int_{-\pi}^{\pi} f(x)\sin nx\,dx \end{cases}$$

"Similarly, in the Fourier integral representation, we can represent $a(\omega)$ and $b(\omega)$ as integrals. We're already using the letter x, so let's use y as the variable of integration."

$$\begin{cases} a(\omega) = \dfrac{1}{\pi}\displaystyle\int_{-\infty}^{\infty} u(y,0)\cos\omega y\,dy \\ b(\omega) = \dfrac{1}{\pi}\displaystyle\int_{-\infty}^{\infty} u(y,0)\sin\omega y\,dy \end{cases}$$

"So by using the Fourier integral, we've found a solution $u(x,t)$ satisfying our initial conditions."

$$\begin{aligned} & u(x,t) \\ &= \int_0^{\infty} e^{-\omega^2 t}\left(a(\omega)\cos\omega x + b(\omega)\sin\omega x\right) d\omega \\ &= \int_0^{\infty} e^{-\omega^2 t}\left(\frac{1}{\pi}\int_{-\infty}^{\infty} u(y,0)\cos\omega y\,dy\cos\omega x + \right. \\ &\qquad \left. \frac{1}{\pi}\int_{-\infty}^{\infty} u(y,0)\sin\omega y\,dy\sin\omega x\right) d\omega \\ &= \frac{1}{\pi}\int_0^{\infty} e^{-\omega^2 t}\int_{-\infty}^{\infty} u(y,0)\left(\cos\omega y\cos\omega x + \sin\omega y\sin\omega x\right) dy\,d\omega \\ &= \frac{1}{\pi}\int_0^{\infty} e^{-\omega^2 t}\int_{-\infty}^{\infty} u(y,0)\cos\omega(x-y)\,dy\,d\omega \end{aligned}$$

from additive theorem and $\cos\omega(y-x) = \cos\omega(x-y)$

"Let's change the order of integration. There are some conditions on being able to do that, but trust me that they're satisfied here."

$$u(x,t) = \frac{1}{\pi}\int_{-\infty}^{\infty} u(y,0)\int_0^{\infty} e^{-\omega^2 t}\cos\omega(x-y)\,d\omega\,dy \qquad \cdots\heartsuit$$

"So," Miruka said, putting down her pen. "What do we do with this?"

"Don't look at me," Yuri said, immediately backing out.

"Actually, I wanted to hear a little more about how the Fourier series is discrete, but the Fourier integral is continuous," I said. "Does that somehow make this double integral easier to deal with?"

"Something about this reminds me of one of our friends," Tetra said.

"A friend?" Yuri said.

"I can't remember the name, but isn't this familiar to something we've seen before?"

"Oh, I know!" I said. "The Laplace integral!"

"That's the one!"

The Laplace integral (see p. 291)

For a real number a,

$$\int_0^\infty e^{-x^2}\cos 2ax\,dx = \frac{\sqrt{\pi}}{2}e^{-a^2}.$$

"Very well," Miruka said. "So let's use the Laplace integral."

$$u(x,t) = \frac{1}{\pi}\int_{-\infty}^{\infty} u(y,0)\int_0^\infty e^{-\omega^2 t}\cos\omega(x-y)\,d\omega\,dy$$

"It's easier to see the correspondence with the Laplace integral if we clean this up using ν as the variable of integration."

$$\int_0^\infty e^{-\omega^2 t}\cos\omega\,(x-y)\;\;d\omega = ? \qquad \text{the integral we want (var. of int. } \omega)$$

$$\int_0^\infty e^{-\nu^2}\cos 2a\nu\;\;d\nu = \frac{\sqrt{\pi}}{2}e^{-a^2} \qquad \text{Laplace integral (var. of int. } \nu)$$

"If we let $\omega = \dfrac{\nu}{\sqrt{t}}$ and $x - y = 2a\sqrt{t}$, we can apply the Laplace integral."

$$\begin{aligned}
\int_0^\infty e^{-\omega^2 t}\cos\omega\,(x-y)\,d\omega &= \int_0^\infty e^{-(\sqrt{t}\omega)^2}\cos\left(\frac{\nu}{\sqrt{t}}\cdot 2a\sqrt{t}\right)d\omega \\
&= \int_0^\infty e^{-\nu^2}\cos 2a\nu\frac{d\omega}{d\nu}\,d\nu \\
&= \frac{1}{\sqrt{t}}\int_0^\infty e^{-\nu^2}\cos 2a\nu\,d\nu && \frac{d\omega}{d\nu} = \frac{1}{\sqrt{t}} \\
&= \frac{1}{\sqrt{t}}\frac{\sqrt{\pi}}{2}e^{-a^2} && \text{Laplace integral} \\
&= \frac{1}{\sqrt{t}}\frac{\sqrt{\pi}}{2}\exp\left(-\frac{(x-y)^2}{4t}\right) \\
&= \frac{\sqrt{\pi}}{2\sqrt{t}}\exp\left(-\frac{(x-y)^2}{4t}\right) && \cdots\cdots\clubsuit
\end{aligned}$$

"Using this..."

$$\begin{aligned}
u(x,t) &= \frac{1}{\pi}\int_{-\infty}^{\infty} u(y,0) \int_0^{\infty} e^{-\omega^2 t}\cos\omega(x-y)\,d\omega\,dy && \text{from } \heartsuit \text{ on p. 342} \\
&= \frac{1}{\pi}\int_{-\infty}^{\infty} u(y,0)\frac{\sqrt{\pi}}{2\sqrt{t}}\exp\left(-\frac{(x-y)^2}{4t}\right)dy && \text{from } \clubsuit \\
&= \int_{-\infty}^{\infty} u(y,0)\frac{1}{2\sqrt{\pi t}}\exp\left(-\frac{(x-y)^2}{4t}\right)dy \\
&= \int_{-\infty}^{\infty} u(y,0)\,w(x,y,t)\,dy
\end{aligned}$$

"...the $w(x,y,t)$ at the end there is this."

$$w(x,y,t) = \frac{1}{2\sqrt{\pi t}}\exp\left(-\frac{(x-y)^2}{4t}\right)$$

"So finally we've found that the solution $u(x,t)$ we were after looks like this."

$$\begin{cases} u(x,t) = \displaystyle\int_{-\infty}^{\infty} u(y,0)\,w(x,y,t)\,dy \\ w(x,y,t) = \dfrac{1}{2\sqrt{\pi t}}\exp\left(-\dfrac{(x-y)^2}{4t}\right) \end{cases}$$

10.6.6 An Analogous Phenomenon

"Let's take a moment to observe this solution $u(x,t)$ to the heat equation we found," Miruka said.

$$u(x,t) = \int_{-\infty}^{\infty} u(y,0)\,w(x,y,t)\,dy$$

"Using this, $u(y,0)$ represents the starting temperature at position y."

$$u(x,t) = \int_{-\infty}^{\infty} \underbrace{u(y,0)}_{\substack{\text{initial temp.}\\ \text{at pos. } y}}\,w(x,y,t)\,dy$$

"What we're doing here is applying the 'weight' $w(x,y,t)$ to this initial temperature, then moving y to take the integral across the entire length of the wire. In other words, this weighting function

$w(x, y, t)$ controls changes in the temperature distribution. The temperature at position x and time t is related to the initial temperature along the wire. However, there are different weightings. See this

$$\exp\left(-\frac{(x-y)^2}{4t}\right)$$

part in the weighting? This shows that a position y has less effect on a position x as the distance between them increases. We can also see that $w(x, y, t) \to 0$ when $t \to \infty$. So no matter the starting form of the temperature distribution $u(x, 0)$, at some point in the far future everything will even out, making the temperature distribution uniform. If we use the Dirac delta function to represent the initial temperature distribution $u(x, 0)$ as $\delta(x)$, we can represent a heat source as a single point. Then we can explicitly calculate $u(x, t)$ like this."

$$\begin{aligned} u(x,t) &= \int_{-\infty}^{\infty} u(y,0)\, w(x,y,t)\, dy \\ &= \int_{-\infty}^{\infty} \delta(y)\, w(x,y,t)\, dy \\ &= \frac{1}{2\sqrt{\pi t}} \int_{-\infty}^{\infty} \delta(y) \exp\left(-\frac{(x-y)^2}{4t}\right) dy \\ &= \frac{1}{2\sqrt{\pi t}} \exp\left(-\frac{x^2}{4t}\right) \end{aligned}$$

"Now we can vary the t in $u(x, t)$ and graph the resulting changes in the temperature distribution."

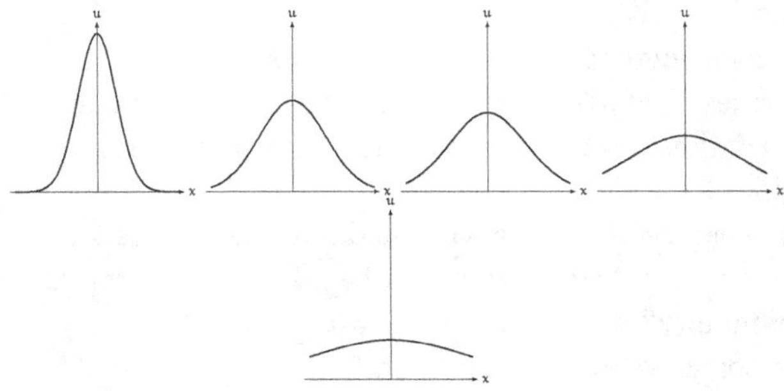

"I'm using it somewhat nonchalantly, but the Dirac delta function $\delta(y)$ isn't a function in the normal sense. It's a generalized function, defined so that the integral of its product with the function $f(y)$ from $-\infty$ to ∞ equal to $f(0)$."

$$\int_{-\infty}^{\infty} \delta(y)\, \underline{f(y)}\, dy = \underline{f(0)} \qquad \text{Dirac } \delta \text{ function}$$

10.6.7 Back to Ricci Flow

"So we've taken a look at Fourier's heat equation as an extremely simplified version of Ricci flow," Miruka said. "We started with the function $u(x, 0)$ as an initial temperature distribution, and considered a continuous transformation of this continuous function by controlling time t. Another way of looking at it is to say we used the heat equation to flatten out an uneven distribution. In principle, Hamilton's Ricci flow does the same when applied to the Poincaré conjecture. By analogy, any initial temperature distribution eventually becoming uniform is similar to any Riemannian metric applied to a closed 3-manifold eventually smoothing out to an even Ricci curvature."

Miruka's speech suddenly slowed.

"In the case of a wire, temperature will always eventually even out. In the case of a closed 3-manifold, however, Ricci curvature doesn't necessarily cooperate so nicely. Specifically, it's possible for singularities that can't be handled as-is to appear. Hamilton used 'surgery' to remove those singularities. So while we took a look at the Fourier heat equation as an analogue of Ricci flow, they're actually quite different. A wire is one-dimensional, and in that case what we're dealing with is temperature, as a real number. There are no Riemannian metrics, or curvature tensors, or Ricci curvatures, or singularities. Unfortunately, we aren't yet at the point where we can tread those paths. We have to be satisfied with these crude analogies—for now."

"For now, right," Tetra said, nodding. "But someday! Someday we'll be able to! And I hope we'll be able to do so together! We'll take on infinity!"

"I'm sorry, we're closing up now."

We turned to see it was a cafeteria employee speaking to us. Looking around, we realized we were the last people in the place. We had covered the table with sheets of paper, filled with mathematics. I felt a sense of déjà vu.

"Guess we'd better head back," I said.

10.7 Us

10.7.1 Past to Future

We left the cafeteria and walked through the school campus. It was already edging toward dark, but we walked slowly, in no hurry.

Miruka, walking ahead of me, turned and said, "So... Get your scores back yet?"

"I did. An 'A,' if only just barely. Wouldn't have been here today otherwise. But let's not bring that up."

"Hmph. Well, congratulations anyway." She gave a shrug and jokingly stuck her tongue out at me.

But her words had snapped me back to reality, invoking an impulsive groan. "Ugh! Next is the real thing!"

"And finals for me!" Tetra said.

"Entrance exams for me too," Yuri said, grabbing onto Tetra's arm.

Tetra brought her face near to Yuri's. "I bet you're looking forward to coming to our high school. You and... him."

"Shhh! That's a secret!"

Huh? What's this?

"I'm heading back to the States soon," Miruka said, looking up at the sky.

"When will we see you again?" Yuri asked.

"We'll see," Miruka replied, but looking at me for some reason, and smiling.

10.7.2 If Winter Comes...

"*Wow* it's cold," Yuri said. "I'd bet winters were cold in Russia, too."

"I like spring breezes, summer breezes, and autumn breezes," I said. "But the winter wind is nothing but cold."

"That's not true," Tetra said. "It's the chill of winter winds that makes our anticipation of spring breezes so joyous! The colder it is, the stronger our hearts yearn for spring!"

"Such a poetic, forward-looking attitude," I said. "By the way, how's *Eulerians* coming along?"

" 'If winter comes, can spring be far behind?' " she quoted. "Lisa and I have gone a long way toward resolving our editorial differences, haven't we?"

"Sure," Lisa said, walking alongside.

"Maybe I can write in *Eulerians* about everything we talked about today, about topology and the Poincaré conjecture and all the mathematicians who contributed to its solution!"

"No," Lisa immediately replied. "Information overload."

"There she goes again . . . "

"Too much for one person, too much for one issue, too much for just you," Lisa continued. "So find more people. Split it between multiple issues. No need to do it all yourself. No need to do it all at once. Divide and conquer."

After this uncharacteristically long pronouncement, Lisa was hit by an extended coughing spell.

"Are you okay?" Tetra asked, rubbing Lisa's back. "You're right, of course. We all have our times for passing the baton."

"Looks like *Eulerians* will last a while," Lisa said.

10.7.3 Can Spring be Far Behind?

"We're heading home already?" Yuri said, a slight whine in her voice. "That's no fun. Maybe we can find a Christmas party or something somewhere!"

"I don't know about you, but I'm done for the day," I said.

"Party pooper."

"Want me to give your cheek another twist?" Miruka joked.

"Ooh! Over there!" Tetra said, pointing at a large tree. "Let's take a group photo! Quick, before it gets dark!"

There was a huge evergreen tree standing near the school gate, so tall I had to crane my neck to see its top. I had no idea what kind of tree it was. I wondered how old it was. We gathered in front of it. Lisa positioned a camera, set its timer, and joined our lineup.

Click.

A moment in time, captured.

If winter comes, can spring be far behind?

I wanted to learn as much as possible at university. I wondered what new people I might meet. What batons I might receive, and who I might pass a baton to. My entrance exams lay right before me. I couldn't see what the future held for me. I could make no predictions about what was to come. I couldn't even be sure I would pass my exams. All I could do was head toward whatever my destination was, one careful step at a time. That was all any of us could do.

Winter had come, so spring couldn't be far behind.

> The trumpet of a prophecy! O Wind,
> If Winter comes, can Spring be far behind?
>
> ---
>
> PERCY BYSSHE SHELLEY
> "Ode to the West Wind"

Epilogue

"Is this you?" the girl said, entering the teachers' lounge with a photograph in her hand.

"Wow, this brings back some memories. Where'd you find it?"

"So it *is* you. It was between the pages of some old newsletter kinda thing I found when we were cleaning the clubroom. You're so *young* ..."

"That was taken in my last year of high school, just before college entrance exams."

"Huh, it's hard to imagine you cramming for exams. It feels like you've been teaching here for eternity."

"Now that's just silly."

"A group photo surrounded by all these girls, just before exams? Very popular with the ladies, weren't you?" She giggled.

"At the time, I had a lot more to be worried about than girls."

"Also hard to imagine."

"C'mon, I'm only human. Everyone has worries."

"Glad I have company."

"What, the Math Club president, not to mention the student with the best grades in her class, has worries?"

"Not a very teacherly thing to say. And yes, entrance exams have got me down, too. I'm so depressed about them I want to cry."

"Spring is on its way."

"Not fast enough. It snowed this morning!"

"That reminds me of a classical Japanese poem by Kiyohara-no-Fukuyabu."

> Still winter lingers,
> but from the heavens fall these
> blossoms of purest white.
>
> It seems that spring must wait
> on the far side of those clouds.

"Falling blossoms? Is he talking about snow?"

"That's right. He's comparing snow in winter to the cherry blossoms in spring. Snowflakes as an analogue for cherry blossoms. Even in the Heian era, a thousand years ago, during winter people longed for spring. That emotion is invariant. The colder it is, the stronger our yearning for warmer days. Huh... If 'yearning for spring' is proportional to the temperature difference, we might be able to set up a differential equation that—"

"Give it a break, teach. I'm just wondering if spring will come at all."

"You've done everything you need to prepare. All that's left is to show them what you're capable of, just like you did on your practice exams."

"I don't know... For some reason, having solved past problems doesn't instill me with confidence for the future."

"That's because just getting a correct answer isn't enough. You have to use those problems as feedback to measure how you're doing, to look for ways you can improve your technique."

"Sure, I'm doing that."

"The same's true even in cutting-edge mathematics. You aren't done when you've found an answer. You have to think about how you solved the problem, about what new problems might lie beyond your solution. That's feedback for the whole world."

"Feedback for the world?"

"When you've newly solved a problem, it's your responsibility to create a new problem. After all, you're now the one who understands it best. Those standing at the forefront can best describe the landscape beyond, so it's your responsibility to do so."

"I'll worry about that when the time comes."

Several students were milling about in the hallway outside the teachers' lounge, peeking in from time to time.

"Seems there's a math club in search of its leader," he said.

"Oops! You're right. Gotta scoot! Later!"

"Yep, see you later."

She left the lounge with a wave, merging with the Math Club crowd. They walked away, cheerfully chatting about their upcoming exams and other things.

He looked out the window, up at the winter sky beyond, feeling the cold and yearning for spring.

It was winter now, but spring could not be far behind.

Still winter lingers,
but from the heavens fall these
blossoms of purest white.
It seems that spring must wait
on the far side of those clouds.

KIYOHARA-NO-FUKAYABU
"On seeing falling snow"
Trans. by L.R. Rodd and M.C. Henkenius

Afterword

This is Hiroshi Yuki, thanking you for reading *Math Girls 6: The Poincaré Conjecture.* This is the sixth volume in the *Math Girls* series, following *Math Girls* (2007), *Math Girls 2: Fermat's Last Theorem* (2008), *Math Girls 3: Gödel's Incompleteness Theorems* (2009), *Math Girls 4: Randomized Algorithms* (2011), and *Math Girls 5: Galois Theory* (2012). As in the previous books, this is a story in which Miruka, Tetra, Yuri, Lisa, and the narrator explore mathematics and youth.

There was a six-year gap between publication of this book and *Math Girls 5*, due largely to the time it took me to wrap my head around the background and assertions of the Poincaré conjecture.

The main mathematical concepts presented in this book include topology, fundamental groups, non-Euclidean geometries, differential equations, manifolds, the Fourier series, and of course the Poincaré conjecture. If you would like to look deeper into these topics, the Recommended Reading section suggests some books you might want to read.

Alongside the *Math Girls* series, I am also writing the *Math Girls Talk About...* series, which covers more basic topics in mathematics. I would like to thank the Mathematical Society of Japan for awarding me their 2014 Publishing Prize in recognition of the *Math Girls* series and my book *Sūgaku bunshō sakuhō* [The Style of Mathematical Writing].

As with the other books in the *Math Girls* series, this book was created using LaTeX 2ε and the AMS Euler font. Also as before, Haruhiko Okumura's book *Introduction to Creating Beautiful Documents with LaTeX 2ε* was an invaluable aid during layout, and I thank him for it. I created the diagrams using OmniGraffle, Ti*k*Z, and TeX2img, and thank the authors of those tools.

I would also like to thank the following persons (and those remaining anonymous) for proofreading and giving me invaluable feedback while I was writing this book. Of course, any remaining mathematical errors are solely the responsibility of the author.

> Ryo Akazawa, Yusuke Ikawa, Haruka Ishii, Tetsuya Ishiu, Kazuhiro Inaba, Ryuhei Uehara, Yakimi Uematsu, Taiki Uchida, Yoichi Uchida, Kento Onishi, Hiromichi Kagami, Takumi Kitagawa, Natsumi Kikuchi, Iwao Kimura, Koki Kirishima, Jun Kudo (@math_neko), Kazuhiro Kezuka, Hiroshi Fujita, Yutori Bonten (Medaka College), Masahide Maehara, Nami Masuda, Atsushi Matsuura, Yoshihiro Matsumori, Kiyoshi Miyake, Ken Murai, Taiju Yamada, Takashi Yoneuchi

I thank my readers for their support of this series, as well as my editor, Kimio Nozawa, for his continuous support during the long process of creating these books.

I thank my dearest wife and my two sons.

I dedicate this book to my mother-in-law, who passed away last year.

Finally, thank *you* for reading my book. I hope that someday, somewhere, our paths shall cross again.

Hiroshi Yuki
2018
https://www.hyuki.com/girl/

Recommended Reading

> The thirteenth-century scholar Ibn Jama'a, though recommending that students purchase books whenever possible, thought it most important that they be "carried in the heart" and not merely kept on a shelf.
>
> ALBERTO MANGUEL
> *The Library at Night*

TO THOSE WHO WILL CONTINUE STUDYING

One of the topics appearing in this book is fundamental groups (the first homotopy group). When reading books on topology, you will also come across mention of "homology groups," but please note that these are different things. I remember that when I first started reading books on topology I mistakenly thought them to be the same, leaving me perplexed as to why different books treated them so differently. If you will pursue further study of topology, please be careful not to make the mistake I did.

[Note: The following references include all items that were listed in the original Japanese version of *Math Girls 6: The Poincaré Conjecture*. Most of those references were to Japanese sources. Where an English version of a reference exists, it is included in the entry.]

General Reading

[1] Negami, S. (2007). *Toporojikaru uchū kanzenban—Poankare-yosō kaiketsu e no michi* [The Topological Universe, Complete Edition—Toward a Solution of the Poincaré Conjecture]. Nihon Hyoronsha.

> An easy-to-read book about topology. I particularly like the author's imagery, such as use of a "space globe" for understanding manifolds and depicting a knife cutting up and expanding a three-dimensional sphere. I referred to this book while writing Chapter 5.

[2] Seyama, S. (2009). *Hajimete no toporojī* [An Introduction to Topology]. PHP Institute.

> A fun read providing an overview of topology. I referred to this book while writing Chapter 2.

[3] Szpiro, G. (2008). *Poincaré's Prize.* Translated by Kajihara, T., Sakai, H., Shiohara, M., and Matsui, N. as *Poankare-yosō* (Hayakawa, 2011).

> A book describing the evolution of modern geometry as driven by the Poincaré conjecture and the challenges faced by the many mathematicians who worked toward its solution.

[4] O'Shea, D. (2011). *The Poincaré Conjecture: In Search of the Shape of the Universe.* Translated by Itokawa, H. as *Poankare-yosō o hodoita sūgakusha* (Nikkei BP, 2007).

> This book presents mathematicians and the math they studied, focusing on the development of geometry, changes in our perception of the shape of the universe, and changes in the social situation surrounding mathematics.

[5] Kasuga, M. (2011). *Hyakunen no nanmon wa naze toketa no ka—Tensai sūgakusha no hikari to kage* [How a Century-old Problem was Solved—The Light and Shadow of Genius Mathematicians]. Shinchosha.

> A book based on an NHK television show of the same name. Introduces many of the mathematicians who worked on a solution to the Poincaré conjecture.

[6] Gessen, M. (2009). *Perfect Rigor: A Genius and the Mathematical Breakthrough of the Century* Translated by Aoki, K. as *Kanzen-naru shōmei* (Bungeishunju, 2009).

A book describing the life of Perelman, including episodes from the International Mathematical Olympiad and other social situations.

[7] Ahara, H. (2017). *Parikore de sūgaku o* [Math at Paris Fashion Week]. Nihon Hyoronsha.

A book written in the form of a dialogue concerning "eight pictures of space" drawn by Thurston himself. Also features numerous photographs from the Issey Miyake Fall 2010 collection "Poincaré Odyssey," which adopted a topological motif.

[8] Negami, S. (ed.) (2011). *Sūgaku seminā zōkan: Tanoshimō! Sūgaku* [Mathematics Seminar, Special Edition: Let's Enjoy Mathematics]. Nihon Hyoronsha.

A book presenting various fun topics from mathematics, including topology and the Poincaré conjecture.

[9] Nihon Hyoronsha (2010). *Sūgaku seminā zōkan: Mireniamushō mondai* [Mathematics Seminar, Special Edition: The Millennium Prize Problems].

A collection of articles related to the Millennium Prize Problems.

The Seven Bridges of Königsberg Problem

[10] Euler, L. (1736). "Solutio problematis ad geometriam situs pertinentis" [The solution of a problem relating to the geometry of position], *Commentarii academiae scientiarum Petropolitanae 8*. pp. 128–140.

The original paper (in Latin) in which Euler presented the bridges of Königsberg problem. I translated the epigraph for Chapter 1 from the following English translations:

- Newman, J. (ed.), *The World of Mathematics, Vol. 1*, pp. 573–580. George Allen & Unwin, 1956.

· Biggs, N., Lloyd, E. and Wilson, R., *Graph Theory, 1736–1936*, pp. 1–11. Clarendon Press, Oxford, 1976.

[11] Hopkins, B. and Wilson, R. (2004). "The Truth about Königsberg," *The College Mathematics Journal*, Vol. 35, pp. 198–207.

A paper about solving the bridges of Königsberg problem as presented in Euler's original paper, describing what Euler did and did not do.

TOPOLOGY

[12] Matsuzaka, K. (1968). *Shūgō isō nyūmon* [An Introduction to Sets and Topologies]. Iwatani Shoten.

A textbook regarding set theory and topology.

[13] Shiga, H. (1988). *Isō e no sanjū-kō* [Thirty Lectures on Topology]. Asakura Shoten.

An easy-to-read, step-by-step guide to learning about metric spaces, topological spaces, compact spaces, complete metric spaces, and other topics in topology.

[14] Kotake, Y., Seyama, S., Tamano, K., Negami, S., Fukaishi, H., and Murakami, H. (1996). *Toporojī mangekyō I* [A Kaleidoscope of Topology, Vol. I]. Asakura Shoten.

The field of topology as viewed from six perspectives: distance spaces, homology theory, knot theory, topological spaces, homotopy theory, and graph theory in topological geometries. I referenced this book while writing Chapter 2.

[15] Seyama, S. (2003). *Toporojī: Yawarakai kikagaku (zōho-ban)* [Topology: The Soft Geometry (Expanded Edition)]. Nihon Hyoronsha.

A "soft" mathematics book covering topics such as the bridges of Königsberg problem, classification of closed surfaces, homology theory, and homology groups. I referenced this book while writing Chapter 1.

[16] Ichiraku, S. (1993). *Isō kikagaku: Shin sūgaku kōza 8* [Topology: New Lectures on Mathematics 8]. Springer.

A textbook on topological geometry that deals with topics such as topological spaces, fundamental groups, covering spaces, the Jordan curve theorem, classification of closed surfaces, and homology groups.

[17] Tamura, I. (1972). *Toporojī* [Topology]. Iwanami Shoten.

A mathematics book on topological geometry covering topics such as topological figures, homology groups, and fundamental groups. Also describes lens spaces and dodecahedral space (the Poincaré homology sphere).

[18] Kojima, S. (1998). *Toporojī nyūmon* [An Introduction to Topology]. Kyoritsu Shuppan.

A textbook about homotopy, Riemann surfaces, fundamental groups, covering spaces, homology, and cohomology.

[19] Ahara, H. (2013). *Keisan de mi ni tsuku toporojī* [Learning Topology through Calculation]. Kyoritsu Shuppan.

A textbook that focuses on classification theorems for homology groups and curved surfaces. I referenced this book while writing Chapter 2.

[20] Ota, H. (2016). *Tanoshimō shaei heimen* [Fun with Projective Planes]. Nihon Hyoronsha.

A textbook that focuses on classification theorems for homology groups and curved surfaces. I referenced this book while writing Chapter 2.

Theory of Curved Surfaces and Manifolds

[21] Coxeter, H. (2009). *Introduction to Geometry.* Translated by Kinbayashi, K. as *Kikagaku nyūmon (Vol. 1)* (Chikuma Shobo, 2015).

A mathematics book covering topics such as projective geometry, hyperbolic geometry, differential geometry of curves and surfaces, the Gauss–Bonnet theorem, and measures of curvature. I referenced this book while writing Chapter 8. [Note: This book is a single volume in English, translated as two volumes in Japanese.]

[22] Umehara, M. and Yamada, K. (2009). *Kyokusen to kyokumen (kaichōban)* [Curves and Surfaces (Revised Ed.)]. Shokabo.

A textbook on curved surface theory from the viewpoints of curves, curved surfaces, and manifolds. I referenced this book while writing about spherical geometry and finding the area of spherical triangles in Chapter 8.

[23] "*Sūri kagaku tokushū: Gausu*" ["Special Edition on the Mathematical Sciences: Gauss"], December 2017 issue. Saiensu-sha (2017).

A special edition feature on the wide-ranging activities of Gauss. I referenced this when writing Chapter 8.

[24] Terasaka, H. and Shizuma, R. (2015). *Jūkyūseiki no sūgaku: Kikagaku II (Sūgaku no rekishi 8-b)* ["Mathematics of the Nineteenth Century: Geometry II (History of Mathematics 8-b)"], Kyoritsu Shuppan.

Translations of Gauss's theory of curved surfaces. I referenced this when writing Chapter 8.

[25] Riemann, B. (1854). "On the Hypotheses which Lie at the Bases of Geometry." Translated by Sugawara, M. as *Kikagaku no kiso o nasu kasetsu ni tsuite* (Chikuma Shobo, 2013).

Riemann's 1854 inaugural presentation.

Euclidean and Non-Euclidean Geometries

[26] Euclid. *Elements.* Translation and commentary by Nakamura, K., Terasaka, H., Ito, S., and Ikeda, M. as *Yūkuriddo genron* (Kyoritsu Shuppan, 2011).

A translation of Euclid's *Elements.* I referenced this while writing Chapter 4.

[27] Kobayashi, S. (1990). *Yūkuriddo kika kara gendai kika e* [From Euclidean Geometry to Modern Geometry]. Nihon Hyoronsha.

This book describes Euclidean geometry and non-Euclidean geometry (the Poincaré half-plane model and Klein model as models of hyperbolic geometry) from the viewpoint of Riemannian geometry.

[28] Ahara, K. (2016). *Sakuzu de mi ni tsuku sōkyoku kikagaku* [Learn Hyperbolic Geometry by Drawing]. Kyoritsu Shuppan.

A book that teaches hyperbolic geometry while using the GeoGebra software program to create concrete diagrams.

[29] Fukaya, K. (2004). *Sōkyoku kika* [Hyperbolic Geometry]. Iwanami Shoten.

A mathematics book that focuses on hyperbolic geometry, one form of non-Euclidean geometry.

[30] Tsuchihashi, H. and Uchida, R. (2017). *Sōkyoku heimen-jō no kikagaku* [Geometry on Hyperbolic Planes]. Iwanami Shoten.

A mathematics book that investigates whether Desargues's theorem and Pascal's theorem holds on the hyperbolic plane. Includes many figures using Poincare's disk model.

[31] Coxeter, H. (2009). *Introduction to Geometry.* Translated by Kinbayashi, K. as *Kikagaku nyūmon (Vol. 2)* (Chikuma Shobo, 2015).

A textbook that depicts geometry through transformation groups. Part I describes isometric transformations and similarity transformations on Euclidean planes and Euclidean spaces, using groups to examine Escher's paintings. [Note: This book is a single volume in English, translated as two volumes in Japanese.]

[32] Ito, T. (2017). "Hyperbolic Non-Euclidean World and Figure-8 Knots". `http://web1.kcn.jp/hp28ah77/`

A website that describes non-Euclidean geometry with many diagrams and computer-generated graphics.

Fourier Series, the Heat Equation, and Partial Derivatives

[33] Shiga, H. (1988). *Sūgaku ga sodatte iku monogatari dai sanshū: Sekibun no sekai (ichiyō shūsoku to fūrie kyūsū)* [Stories Developed by Mathematics, Week 3: The World of Integrals (Uniform Convergence and the Fourier Series)]. Iwanami Shoten.

An easy-to-read description of Fourier series. I referenced this book while writing Chapter 9.

[34] Kogure, Y. (1999). *Nattokusuru fūrie henkan ichiyō shūsoku* [Understandable Uniform Convergence of Fourier Series]. Kodansha.

A textbook for learning Fourier series, Fourier transforms, and Fourier analysis through concrete calculations. I referenced this book while writing Chapters 9 and 10.

[35] Gowers, T., Barrow-Green, J., and Leader, I. (Eds.). (2008). *The Princeton Companion to Mathematics.* Translated by Sunada, T., Hirata, N., Futaki, A., and Mori, M. as *Purinsuton sūgaku taizen* (Asakura Shoten, 2015).

A compendium of mathematics presenting topics from varied viewpoints. I referenced this book while writing about the analogue for Ricci flow in Chapter 10.

[36] Maeno, M. (2016). *Bijuaru gaido butsuri sūgaku: Ichi-hensū no bisekibun to jōbibun hōteishiki* [Visual Guide to Physical Mathematics: Univariate Calculus and Ordinary Differential Equations]. Tokyo Tosho.

A mathematics book that uses many figures to clearly explain calculus and ordinary differential equations.

[37] Maeno, M. (2017). *Bijuaru gaido butsuri sūgaku: Tahensūkansū to hen-bibun* [Visual Guide to Physical Mathematics: Multivariate Calculus and Partial Derivatives]. Tokyo Tosho.

A mathematics book that uses many figures to clearly explain multivariate calculus and partial derivatives.

[38] Tasaki, H. (2020). *Sūgaku: Butsuri o manabi tanoshimu tame ni* ["Mathematics for Learning and Enjoying Physics"]. `http://www.gakushuin.ac.jp/~881791/mathbook/`

A math textbook written for those studying physics and related fields. A draft of this book is published as a PDF at the above URL. I primarily referenced this book while writing about differential equations.

The Poincaré conjecture

[39] Saito, T. (Trans.), Adachi, T., Sugiura, M., and Nagaoka, R. (Eds.) (1996). *Poankare toporojī* [Poincaré's Topology]. Asakura Shoten.

Translations of four papers written by Poincaré regarding his conjecture. An appendix provides commentary by Yukio Matsumoto regarding the basic conjectures of topology, polygon triangulation, and the Poincaré conjecture.

[40] *Sūgaku seminā zōkan: Kaiketsu! Poankare-yosō* ["Mathematics Seminar Special Edition: The Poincaré Conjecture, Solved!"]. (2007). Nihon Hyoronsha.

A collection of articles from the magazine *Mathematics Seminar* regarding the Poincaré Conjecture. Articles by many authors concerning the Poincaré conjecture, Thurston's geometrization conjecture, Hamilton's Ricci flow, the methods used in Perelman's proof, and other topics.

[41] Monastyrsky, M. (1998). *Modern Mathematics in the Light of the Fields Medals.* Translated by Mano, G. as *Fīruzu-shō de miru gendai sūgaku* (Chikusho Shobo, 2013).

An analysis of modern mathematics with a focus on works for which the Fields Medal was awarded. Provides a simple description of Perelman's achievements.

[42] Smale, S. "Generalized Poincare's conjecture in dimensions greater than four" (1961). *The Annals of Mathematics, 2nd Ser.*, Vol. 74, No. 2. Princeton University.

Smale's paper giving a proof of the Poincaré Conjecture in higher dimensions.

[43] Freedman, M. "The topology of four-dimensional manifolds" (1982). *Journal of Differential Geometry, Vol. 17, No. 3.* International Press.

Friedman's paper giving a proof of the Poincaré Conjecture in four dimensions.

[44] Kobayashi, J. (2011). *Ricchi-furō to kikakayosō* [Ricci Flow and the Geometrization Conjecture]. Baifukan.

A technical book that describes in detail how Hamilton and Perelman solved Thurston's geometrization conjecture.

[45] International Mathematical Union. (2006). "For his contributions to geometry and his revolutionary insights into the analytical and geometric structure of the Ricci flow" [Press release]. Retrieved from https://www.mathunion.org/fileadmin/IMU/Prizes/Fields/2006/PerelmanENG.pdf.

A press release describing why Perelman was awarded the Fields Medal.

[46] Clay Mathematics Institute. (2010). "Prize for Resolution of the Poincaré Conjecture Awarded to Dr. Grigori Perelman" [Press release]. Retrieved from https://claymath.org/sites/default/files/millenniumprizefull.pdf.

A press release by the Clay Mathematics Institute, sponsor of the Millennium Prize Awards, regarding Perelman's solution of the Poincaré conjecture.

[47] Kobayashi, J. (2012). Poankare-yosō wa ikani shite kaiketsu sareta ka ["How the Poincaré Conjecture was Solved"] [Website]. Retrieved from https://www.math.nagoya-u.ac.jp/ja/public/2012/download/homecoming2012_kobayashi.pdf.

A slideshow describing Poincaré's conjecture and its solution, starting from Gauss's theory of intrinsic surfaces, Riemannian geometry, and Klein's geometry.

Perelman's Papers

[48] Perelman, G. (2002) "The entropy formula for the Ricci flow and its geometric applications" (available at http://arxiv.org/abs/1904.09927).

[49] Perelman, G. (2003) "Ricci flow with surgery on three-manifolds" (available at https://arxiv.org/abs/math/0303109).

[50] Perelman, G. (2003) "Finite extinction time for the solutions to the Ricci flow on certain three-manifolds" (available at https://arxiv.org/abs/math/0307245).

THE *Math Girls* SERIES

[51] Yuki, H. (2011). *Math Girls.* Bento Books. Published in Japan as *Sūgaku gāru* (Softbank Creative, 2007).

The story of two girls and a boy who meet in high school and work together after school on mathematics unlike anything they find in class. In the school library, classrooms, and cafes, they investigate topics like prime numbers, absolute values, the Fibonacci sequence, the relation between arithmetic and geometric means, convolutions, harmonic numbers, the zeta function, Taylor expansions, generating functions, the binomial theorem, Catalan numbers, and numbers of partitions.

[52] Yuki, H. (2012). *Math Girls 2: Fermat's Last Theorem.* Bento Books. Published in Japan as *Sūgaku gāru / Ferumā no saishū teiri* (Softbank Creative, 2008).

The second book in the *Math Girls* series, where new "math girl" Yuri joins the others on a quest to understand "the true form" of the integers. Presents groups, rings, and fields among other topics, building to a tour of Fermat's last theorem. Other topics include prime numbers, the Pythagorean theorem, Pythagorean triples, prime factorization, greatest common divisors, least common multiples, proof by contradiction, the pigeonhole principle, the definition of groups, abelian groups, integer remainders, congruence, and Euler's formula.

[53] Yuki, H. (2016). *Math Girls 3: Gödel's Incompleteness Theorems.* Bento Books. Published in Japan as *Sūgaku gāru / Gēderu no saishūteiri* (Softbank Creative, 2009).

The third book in the *Math Girls* series, where the math girls (and boy) use formal systems to learn "the mathematics of mathematics" and learn about Gödel's oft-misunderstood theorems. Topics include the Peano axioms, mathematical induction, basic set theory, Russell's paradox, mappings, limits, why $0.999\cdots = 1$, the basics of mathematical logic, the ϵ–δ definition of limits, diagonalization, equivalence relations, radians, the sine and cosine functions, Hilbert's program, and a proof of Gödel's incompleteness theorems.

[54] Yuki, H. (2019). *Math Girls 4: Randomized Algorithms.* Bento Books. Published in Japan as *Sūgaku gāru / Rantaku arugorizumu* (Softbank Creative, 2011).

The fourth book in the *Math Girls* series, in which everyone moves ahead one grade in school and are joined by computer whiz Lisa. Together, they explore applications of probability to create algorithms that employ randomization and learn how to quantitatively analyze algorithms. Other topics include the Monty Hall problem, permutations and combinations, Pascal's triangle, the definition of probability, sample spaces, probability distributions, random variables, expected values, indicator random variables, order, big-O notation, matrices, linear transformations, matrix diagonalization, random walks, the 3-SAT problem, the P versus NP problem, linear searches, binary searches, and algorithms for bubble sort, quicksort, and randomized quicksort.

[55] Yuki, H. (2020). *Math Girls 5: Galois Theory.* Bento Books. Published in Japan as *Sūgaku gāru / Garoa riron* (Softbank Creative, 2012).

The fifth book in the *Math Girls* series, in which our characters learn the basics of group theory and modern algebra, as initiated by the young Galois on his deathbed. Topics covered include *amidakuji* ("ladder networks"), formulas for solving equations, relations between solutions and coefficients, angle trisections, symmetry formulas, straightedge and compass construction, vector spaces, Lagrange resolvents, Cayley graphs, groups and fields, abelian groups, cyclic groups, symmetric groups, solvable groups, normal subgroups, Lagrange's theorem, minimal polynomials, irreducible polynomials, extension fields, and the Galois correspondence.

[56] Yuki, H.: *Math Girls* `http://www.hyuki.com/girl/en.html`.

The English version of the author's *Math Girls* web site, listing titles in and news related to the *Math Girls* series.

"Obviously, she knew so much because she studied hard."

Hiroshi Yuki
Math Girls 6: The Poincaré Conjecture

Index

Other works by Hiroshi Yuki

(in English)

- *Math Girls*, Bento Books, 2011
- *Math Girls 2: Fermat's Last Theorem*, Bento Books, 2012
- *Math Girls 3: Gödel's Incompleteness Theorems*, Bento Books, 2016
- *Math Girls 4: Randomized Algorithms*, Bento Books, 2020
- *Math Girls 5: Galois Theory*, Bento Books, 2021
- *Math Girls Manga, Vol. 1*, Bento Books, 2013
- *Math Girls Manga, Vol.* 2, Bento Books, 2016
- *Math Girls Talk About Equations & Graphs*, Bento Books, 2014
- *Math Girls Talk About the Integers*, Bento Books, 2014
- *Math Girls Talk About Trigonometry*, Bento Books, 2014

(in Japanese)

- *The Essence of C Programming*, Softbank, 1993 (revised 1996)

- *C Programming Lessons, Introduction*, Softbank, 1994 (Second edition, 1998)
- *C Programming Lessons, Grammar*, Softbank, 1995
- *An Introduction to CGI with Perl, Basics*, Softbank Publishing, 1998
- *An Introduction to CGI with Perl, Applications*, Softbank Publishing, 1998
- *Java Programming Lessons (Vols. I & II)*, Softbank Publishing, 1999 (revised 2003)
- *Perl Programming Lessons, Basics*, Softbank Publishing, 2001
- *Learning Design Patterns with Java*, Softbank Publishing, 2001 (revised and expanded, 2004)
- *Learning Design Patterns with Java, Multithreading Edition*, Softbank Publishing, 2002
- *Hiroshi Yuki's Perl Quizzes*, Softbank Publishing, 2002
- *Introduction to Cryptography Technology*, Softbank Publishing, 2003
- *Hiroshi Yuki's Introduction to Wikis*, Impress, 2004
- *Math for Programmers*, Softbank Publishing, 2005
- *Java Programming Lessons, Revised and Expanded (Vols. I & II)*, Softbank Creative, 2005
- *Learning Design Patterns with Java, Multithreading Edition, Revised Second Edition*, Softbank Creative, 2006
- *Revised C Programming Lessons, Introduction*, Softbank Creative, 2006
- *Revised C Programming Lessons, Grammar*, Softbank Creative, 2006

- *Revised Perl Programming Lessons, Basics*, Softbank Creative, 2006
- *Introduction to Refactoring with Java*, Softbank Creative, 2007
- *Math Girls / Fermat's Last Theorem*, Softbank Creative, 2008
- *Revised Introduction to Cryptography Technology*, Softbank Creative, 2008
- *Math Girls Comic (Vols. I & II)*, Media Factory, 2009
- *Math Girls / Gödel's Incompleteness Theorems*, Softbank Creative, 2009
- *Math Girls / Randomized Algorithms*, Softbank Creative, 2011
- *Math Girls / Galois Theory*, Softbank Creative, 2012
- *Math Girls / The Poincaré conjecture*, Softbank Creative, 2012
- *Java Programming Lessons, Third Edition (Vols. I & II)*, Softbank Creative, 2012
- *Etiquette in Writing Mathematical Statements: Fundamentals*, Chikuma Shobo, 2013
- *Math Girls Secret Notebook / Equations & Graphs*, Softbank Creative, 2013
- *Math Girls Secret Notebook / Let's Play with the Integers*, Softbank Creative, 2013
- *The Birth of Math Girls*, Softbank Creative, 2013
- *Math Girls Secret Notebook / Round Trigonometric Functions*, Softbank Creative, 2014

- *Mathematical Writing, Refinement Edition*, Chikuma Shobo, 2014
- *Math Girls Secret Notebook / Chasing Derivatives*, Softbank Creative, 2015

CPSIA information can be obtained
at www.ICGtesting.com
Printed in the USA
LVHW100153100123
736836LV00002B/5

9 781939 326508